VI School on Cosmic Rays and Astrophysics 2015

AA006008

Journal of Physics: Conference Series Volume 866

Chiapas, Mexico
17 - 25 November 2015

Editors:

Oscar Saavedra San Martin
Karen Salome Caballero Mora

ISBN: 978-1-5108-4421-6
ISSN: 1742-6588

Printed from e-media with permission by:

Curran Associates, Inc.
57 Morehouse Lane
Red Hook, NY 12571

Printed by Curran Associates, Inc. (2017)

For permission requests, please contact the Institute of Physics at the address below.

Institute of Physics
Dirac House, Temple Back
Bristol BS1 6BE UK

Phone: 44 1 17 929 7481
Fax: 44 1 17 920 0979

techtracking@iop.org

Additional copies of this publication are available from:

Curran Associates, Inc.
57 Morehouse Lane
Red Hook, NY 12571 USA
Phone: 845-758-0400
Fax: 845-758-2633
Email: curran@proceedings.com
Web: www.proceedings.com

Table of contents

Volume 866

VI School on Cosmic Rays and Astrophysics
17–25 November 2015, Chiapas, Mexico

Accepted papers received: 20 June 2017
Published online: 5 July 2017

Preface

Papers

IOP Conf. Series: Journal of Physics: Conf. Series **866** (2017) 011001 doi:10.1088/1742-6596/866/1/011001

VI School on Cosmic Rays and Astrophysics 17-25 November 2015, Chiapas, Mexico

The VI School on Cosmic Rays and Astrophysics was held at the MCTP, at the Autonomous University of Chiapas (UNACH), Tuxtla Gutiérrez, Chiapas, Mexico thanks to the Science for Development ICTP-UNACH-UNESCO Regional Seminar, 17-25 November 2015 (http://mctp.mx/e_VI_School_on_Cosmic_Rays_and_Astrophysics.html).
The School series started in La Paz, Bolivia in 2004 and it has been, since then, hosted by several Latin American countires: 1.- La Paz, Bolivia (August, 2004), 2.- Puebla, Mexico (September, 2006), 3.- Arequipa, Peru (September, 2008), 4.- Santo André, Brazil (September, 2010), 5.- La Paz, Bolivia (August, 2012). It aims to promote Cosmic Ray (CR) Physics and Astrophysics in the Latin American community and to provide a general overview of theoretical and experimental issues on these topics. It is directed to undergraduates, postgraduates and active researchers in the field. The lectures introduce fundamental Cosmic Ray Physics and Astrophysics with a review of standards of the field. It is expected the school continues happening during the next years following a tradition.
In this edition, the list of seminars included topics such as experimental techniques of CR detection, development of CR showers and hadronic interactions, composition and energy spectrum of primary CR, Gamma-Ray Bursts (GRBs), neutrino Astrophysics, spacecraft detectors, simulations, solar modulation, and the current state of development and results of several astroparticle physics experiments such as The Pierre Auger Observatory in Argentina, HAWC in Mexico, KASCADE and KASCADE Grande, HESS, IceCube, JEM-EUSO, Fermi-LAT, and others.

This time the school has been complemented with the ICTP-UNACH-UNESCO Seminar of theory on Particle and Astroparticle Physics. The organization was done by MCTP, the Mesoamerican Centre for Theoretical Physics.

The school had 46 participants, 30 students from Honduras, Brazil, Guatemala, Panama, Colombia, Ecuador and Mexico, and 16 lecturers from United States of America, Italy, Japan and Mexico. As a result a total of 17 contributions are included in the present proceedings.

Dr. Oscar Saavedra San Martín
Dr. Karen Salomé Caballero Mora
Editors

IOP Conf. Series: Journal of Physics: Conf. Series **866** (2017) 011001 doi:10.1088/1742-6596/866/1/011001

Participants

IOP Conf. Series: Journal of Physics: Conf. Series **866** (2017) 011002 doi:10.1088/1742-6596/866/1/011002

List of participants of the VI School on Cosmic Rays and Astrophysics 17-25 November 2015, Chiapas, Mexico

Lecturers:

1.- Oscar Saavedra San Martín
Università di Torino and Istituto Nazionale di FisicaNucleare (INFN, Italy)
2.- Mario Rodríguez Cahuantzi
Benemérita Universidad Autónoma de Puebla (BUAP)
3.- Alberto Carramiñana
Instituto Nacional de Astrofísica, Óptica y Electrónica (INAOE, Mexico)
4.- Marco Casolino
Rikagaku Kenkyusho, Research Institute (RIKEN, Japan) and INFN (Italy)
5.- Paolo Desiati
Wisconsin IceCube Particle Astrophysics Center (WIPAC), Department of Astronomy (Madison, USA)
6.- Paolo Lipari
Sezione Roma I dell' INFN, Universita' degli Studi "La Sapienza" c/o Istituto di Fisica (Italy)
7.- Juan Carlos Arteaga Velázquez
Instituto de Física y Matemáticas of Universidad Michoacana de San Nicolás de Hidalgo (IFM-UMSNH, Mexico)
8.- Karen Salomé Caballero Mora
Facultad de Ciencias en Física y Matemáticas, Universidad Autónoma de Chiapas (FCFM-UNACH, Mexico)
9.- Pat Harding
Los Alamos National Laboratory (LANL, USA)
10.- Omar Miranda Romagnoli
Departamento de Física, Centro de Investigación y de Estudios Avanzados del Instituto Politécnico Nacional (CINVESTAV-Mexico)
11.- Pierluigi Belli
INFN Roma Tor Vergata (Italy)
12.- Giuseppe Di Sciascio
Departament of Physics (INFN, Italy)
13.- César Álvarez Ochoa
FCFM-UNACH, Mexico
14.- Luis Villaseñor
IFM-UMSNH, Mexico
15.- Jorge Romo
FCFM-UNACH, Mexico
16.- Lukas Nellen
Instituto de Ciencias Nucleares, Universidad Nacional Autónoma de México (ICN-UNAM, Mexico)

IOP Conf. Series: Journal of Physics: Conf. Series **866** (2017) 011002 doi:10.1088/1742-6596/866/1/011002

Students:

1.- Ibrahim Torres, INAOE, Mexico
2.-Andrea Correa, Escuela Politécnica Nacional, Ecuador
3.- Humberto Martínez Huerta, CINVESTAV, Mexico
4.- Iván Antonio Cárdenas Muñoz, Universidad de Sonora, Mexico
5.- Juan Antonio Salazar Contreras, Escuela de Ciencias Físicas y Matemáticas, Universidad de San Carlos de Guatemala, USAC, Guatemala
6.- Pravda Linda Flor Cabrera Vogel, USAC, Guatemala
7.- Luis Guillermo García Ordóñez, USAC, Guatemala
8.- Marco Antonio Morales Ovalle, USAC, Guatemala
9.- Joshua Raphael Lémus Castillo, USAC, Guatemala
10.- Wendel Alexander Gutiérrez Paxtor, USAC, Guatemala
11.- Julio Roberto Castillo Elías, USAC, Guatemala
12.- Peter Darwin Argueta Ordoñez, USAC, Guatemala
13.- Eddy Brandon de León Aguilar, USAC, Guatemala
14.- Iván Rene Morales Argueta, USAC, Guatemala
15.- Guilherme Tomishiyo, Instituto de Física de São Carlos, Brazil
16.- Rodrigo Guedes Lang, Instituto de Física de São Carlos, Universidade de São Paulo, Brazil
17.- Carlos Eduardo Gabarrete Fajardo, Universidad Nacional Autónoma de Honduras, Honduras
18.- Geovanni Rangel Cortés, UMSNH, Mexico
19.- Pablo Weigandt, Universidad de Panamá, Panama
20.- Andrés Felipe Izquierdo Cartagena, Universidad de Antioquia, Colombia
21.- Rebeca López, BUAP, Mexico
22.- Elsa Alejandra Parra Flores, BUAP, Mexico
23.- Eduardo Moreno Barbosa, BUAP, Mexico
24.- Lenin Pérez Estrada, FCFM-UNACH, Mexico
25.- Laila Vleeschower Calas, FCFM-UNACH, Mexico
26.- Luis Rodolfo Pérez Sánchez, FCFM-UNACH, Mexico
27.- Itzel Amayrani Martínez Salazar, FCFM-UNACH, Mexico
28.- Pedro Alfonso Valencia Esquipula, FCFM-UNACH, Mexico
29.- Hernán Castellanos Valdés, FCFM-UNACH, Mexico
30.- Lizardo Valencia Palomo, FCFM-UNACH

Organizers

Local committee:
Dr. Juan Carlos Arteaga Velázquez

International Committee:
Dra. Karen Salomé Caballero Mora
Dr. Oscar Saavedra San Martín
Dr. Arnulfo Zepeda Domínguez

Acknowledgements

We gratefully acknowledge the kind and efficient help of the organizers Dr. Arnulfo Zepeda Domínguez, Dr. Juan Carlos Arteaga Velázquez, Dr. Humberto Martínez Huerta, M.Sc. Cederik León De León Acuña, and MCTP staff members.

We also thank to the sponsors MCTP, UNACH, Red HAWC and the Mexican National Council for Science and Technology (CONACyT).

Dr. Oscar Saavedra San Martín
Dr. Karen Salomé Caballero Mora
Editors

IOP Conf. Series: Journal of Physics: Conf. Series **866** (2017) 011004 doi:10.1088/1742-6596/866/1/011004

Peer review statement

All papers published in this volume of *Journal of Physics: Conference Series* have been peer reviewed through processes administered by the proceedings Editors. Reviews were conducted by expert referees to the professional and scientific standards expected of a proceedings journal published by IOP Publishing.

IOP Conf. Series: Journal of Physics: Conf. Series **866** (2017) 012001 doi:10.1088/1742-6596/866/1/012001

From The Pierre Auger Observatory to AugerPrime

Alejandra Parra[a] **and Oscar Martínez Bravo**[a], **for the Pierre Auger Collaboration**[b]

[a]Avenida San Claudio y 18 Sur, Colonia San Manuel, Edificio FM1-101B, Ciudad Universitaria, C.P. 72570, Puebla, México.
[b] Observatorio Pierre Auger, Av. San Martin Norte 304, 5613 Malargüe, Argentina (Full author list: *http://www.auger.org/archive/authors_2015_11.html*)

E-mail: `alejandra.parra@alumno.buap.mx`

Abstract. In the present work we report the principal motivation and reasons for the new stage of the Pierre Auger Observatory, AugerPrime. This upgrade has as its principal goal to clarify the origin of the highest energy cosmic rays through improvement in studies of the mass composition. To accomplished this goal, AugerPrime will use air shower universality, which states that extensive air showers can be completely described by three parameters: the primary energy E_0, the atmospheric shower depth of maximum X_{max}, and the number of muons, N_μ. The Auger Collaboration has planned to complement its surface array (SD), based on water-Cherenkov detectors (WCD) with scintillator detectors, calls SSD (Scintillator Surface Detector). These will be placed at the top of each WCD station. The SSD will allow a shower to shower analysis, instead of the statistical analysis that the Observatory has previously done, to determine the mass composition of the primary particle by the electromagnetic to muonic ratio.

1. Introduction

The origin of the ultra-high energy cosmic rays (UHECR) is still an open science case. The UHECR offer information about the most powerful astrophysical sources and they allow one to investigate particle acceleration at the highest energies, in a range that can not be covered by the particle accelerators. The mass composition of ultra-high energy cosmic rays is a very important observable to understand their origin. The atmospheric depth at which the shower reaches its maximum size, X_{max}, is one of the most robust variables to infer the mass composition of air showers.

The Pierre Auger Observatory uses fluorescence detectors to measure the X_{max} and operates with an enhancement, the High Elevation Auger Telescopes (HEAT), which allows Auger to go down in energy to 10^{17} eV, the range of study covers from 10^{17} to $> 10^{19}$ eV.

With three decades of data, the Pierre Auger Observatory has some interesting results about X_{max}. Between 10^{17} and $10^{18.3}$ eV, X_{max} increases by around 85 g cm^{-2} per decade of energy, Fig. 1. This value is larger than the one expected for a constant mass composition ($\sim$ 60 g cm^{-2}/decade) and indicates that the mean primary mass is getting lighter. Around $\approx 10^{18.3}$ eV the observed rate of change of $\langle X_{\mathrm{max}}\rangle$ becomes significantly smaller ($\sim$ 26 g cm^{-2}/decade) indicating that the composition is becoming heavier. The first two moments of the X_{max} distribution ($\langle X_{\mathrm{max}}\rangle$ and $\sigma(X_{\mathrm{max}})$) are related to the first two moments of the distribution of the logarithm of masses of primary particles ($\langle \ln A\rangle$ and $\sigma^2(\ln A)$) as shown in [1]. The lowest values

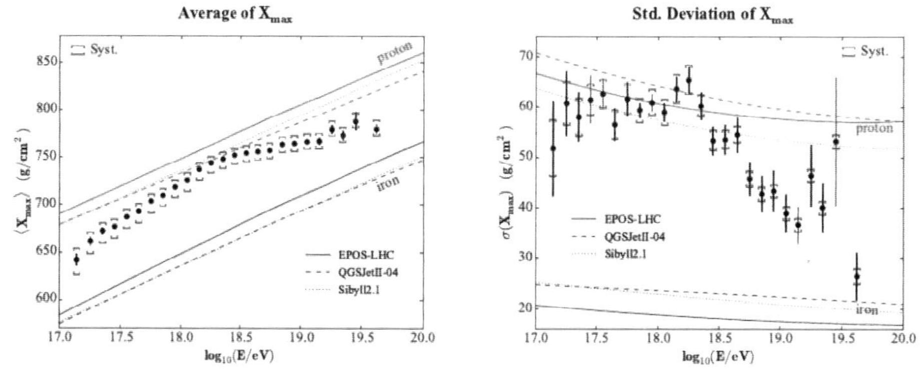

Figure 1. In the left plot the mean of the measured $X_{\max}$ distribution is shown. In the right plot is the standard deviation of the $X_{\max}$ distribution. In both cases the $X_{\max}$ as a function of energy is compared to air-shower simulations for proton and iron.

for the primary mass are at around an energy of $10^{18.3}$ eV, and increase for higher energies. The change might be an indication that the relative fraction of protons becomes smaller for energies above $\approx 10^{18.3}$ eV (see [2]). The implication of the distributions of $X_{\max}$ have been studied in detail by the Pierre Auger Collaboration by considering different assumptions on composition and on hadronic interaction models [2]. Despite use of the interaction models, the Auger data are not well described by a mix of protons and iron nuclei over most of the energy range. The best fits are obtained by considering intermediate masses for the mass composition.

2. Air Shower Universality

Air shower universality says that extensive air showers (EAS) can be completely characterized by three parameters: the primary energy E_0, the atmospheric depth of shower maximum $X_{\max}$, and the number of muons, N_μ. From these parameters, the $X_{\max}$ and the N_μ depend of the mass of the primary particle, and are subject to significant shower to shower fluctuations. It is possible to place constraints on hadronic interaction models, once these two parameters are measured and compared with simulations. In some previous studies it has been proved that the energy spectra and the angular distributions of electromagnetic particles [3], as well as the lateral distribution of energy close to the shower core, are universal [4] [5], i.e. they are functions of E_0, N_μ and the atmospheric depth $X_{\max}$ only.

Another important fact we should consider is that some aspects of Auger SD data are not completely understood, such as: the attenuation curve,the difference between the spectrum obtained from inclined events (which are events with zenith angle between 60° and 80°) in comparison with the spectrum obtained from data with zenith angles less than 60°, when they are shown as a fractional difference (also called the residual) [6]. Also, the information of the FADC traces from the water-Cherenkov stations (belonging to the SD) located far from shower cores have been used to make a reconstruction of the muon production depth, however the current level of systematic uncertainties associated with its determination prevent us from using this to make conclusive statements on mass composition [7].

All the approaches presented so far determine the muon normalization indirectly, leaving it convoluted with the effects of shower development and fluctuations, or relying on the energy scale given by the fluorescence data.

3. Observables

In this work we present the universality in the interpretation of surface detector data. We need to separate the properties of the showers, including the electromagnetic component, the average depth of shower maximum $\langle X_{\max}\rangle$, the normalization of the muon signal at 1000 m (N_μ) and the surface detector energy scale [8]. The normalization of the muon signal is directly related to the number of muons in the shower. The $\langle X_{\max}\rangle$ has been measured by the fluorescence detector with very well known uncertainties. We used this knowledge to infer the number of muons as function of energy in the same way as the energy scale from surface detectors.

The ultra-high energy cosmic rays are very difficult to measure because they have a low rate. For this reason, the Observatory covers a large area, with large separation between detectors. In this way, the shower properties are measured from different points at different distances from the shower core. When the size of the shower is reconstructed, the lateral distribution function (LDF), which describes the fall-off of signal size with the distance from the shower core, leads to a related a uncertainty in both the location of the shower core and in the measurement of the integrated LDF. To avoid the large measurements fluctuations, Hillas proposed a method using the signal at some distance (r_{opt}) from the shower core to classify the size of the shower, and ultimately the energy of the primary particle [9]. The advantages of this method are: (*i*) the effect of uncertainties in the LDF are minimized at a particular core distance, and (*ii*) although the total number of particles at ground level is subject to large fluctuations, the fluctuations of the particle density at the chosen distance from the core are quite small. For example it was shown that at 10^{17} eV the RMS variation in the total number of particles is $\approx 67\%$; for the same shower the RMS variation in the signal of a water-Cherenkov detector at 950 m is $\approx 6\%$. His conclusion was shown to be robust for a variety of hadronic propagation models and energies.

To measure the energy of the primary particle two steps are necessary. First, the detector signal must be measured at some distance from the core and, second, this characteristic signal must be linked to the energy of the primary particle. The Pierre Auger Observatory as a hybrid detector has the advantage of having a fluorescence detector which measures around 10% of the EASs observed with the surface array. The calorimetric energy measurement from the fluorescence detectors can be used to calibrate the characteristic signal from the surface detector. Identifying the optimum core distance r_{opt} for arrays with large spacing is problematic. Identifying the r_{opt} at which to measure the signal of an air shower, will minimize the uncertainty in the energy of the primary particle which follows from a lack of the knowledge of the true LDF. For the Pierre Auger Observatory the r_{opt} is 1000 m, so the characteristic signal will be $S(1000)$ and will be used to determine the energy of the primary particle. Around 1000 m the expected signal is robust against inaccuracies in the assumed LDF, better that 5% [10].

4. Electromagnetic and Muonic Component

In an extensive air shower the electromagnetic component (EM) is initiated by neutral pions, the EM component reaches its atmospheric depth of maximum directly related to the primary mass. The EM component shows universality features that have been used to study the Cherenkov light production in EAS [11] and is a very good estimator of the primary energy.

On other hand, the muon component is due to the charged pions; its maximum is about 100 g cm^{-2} deeper in the atmosphere. The muons bring information on the primary particle mass and the hadronic interaction models. The study of the well known "muon deficit" in simulations [12] as compared to data is mandatory in order to use the muons as composition observables. Also the study of muons can reduce the systematics in the energy scale.

Separating these two components with the Surface Auger for the Pierre Auger Observatory, can be very difficult:

- vertical muons deposit on average 240 MeV in the surface station, the signal being proportional to the station volume, their flux is from 1 to 3 orders of magnitude lower than the EM one.

- the EM component is made up of $\gamma's$ and, roughly one order of magnitude less, by $e^{\pm}$. The surface station acts in this case as a total absorption calorimeter. The average deposited energy per particle is $5 - 10$ MeV.

The temporal profile of the shower can be studied with great detail in each SD station, because their electronics digitizes the signals with a sampling rate of 40 MHz (in AugerPrime will increase to 120 MHz). Thanks to this characteristic it is possible to develop some techniques to separate the signal components [13]. Some of them are: the deconvolution of peaks in traces, the study of jumps and subsequent developments, the smoothing, or methods where the information about muons comes from universality properties of showers [8] [14].

In the current model adopted by the Observatory, shower universality has been tested in terms of the shower plane signal, i.e. there were only considered the shower particles passing through the top of a detector placed perpendicular to the shower axis. This avoids zenith angle dependencies [8]. The study was performed using simulations, the shower plane signals were divided into electromagnetic particles and muons. In order to have a better separation of the two components, there was included the signal from the electromagnetic decay products of muons in the muon component, leaving a "pure" electromagnetic component (S_{em} in VEM). It is important to mentioned that the electromagnetic decay products of muons contributes with 15% of the muon signal in the detector.

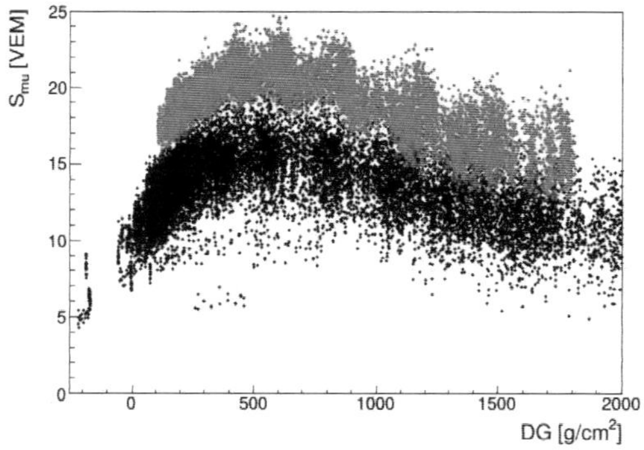

Figure 2. Simulated muon signals, in black for proton and in red for iron. The showers were simulated using QGSJetII/Fluka, between 0° and 70°.

In the Fig. 2, is shown the muon signal (S_μ) as function of DG $= X_{\text{ground}} - X_{\max}$, which is the distance from the shower maximum to the detector along the shower axis (in g/cm^2). The proposed method has shown that the muon and electromagnetic components at 1000 m are completely determine by DG, and the overall muon normalization, N_μ. The parameter DG depends of $X_{\max}$ and θ, and it is possible to do a parametrization of $S_{1000} = S_{1000}(X_{\max}, N_\mu, \theta)$ for a given energy.

5. AugerPrime

AugerPrime will be able to measure the different abundances of muons, photons and electrons in air showers at the surface ground level, to obtain a more precise handle on the primary cosmic ray composition, with increased statistics at the highest energies. The key to differentiating the ground-level air shower particles lies in improving the detection capabilities of the surface array.

AugerPrime consists in covering each of the 1660 water-Cherenkov surface station with planes of plastic-scintillator detectors (SSD) of area 4 m^2. This enhancement will allow the Auger

IOP Conf. Series: Journal of Physics: Conf. Series **866** (2017) 012001 doi:10.1088/1742-6596/866/1/012001

Observatory to determine the electron/photon versus muon abundances of air showers. The scintillator detectors will be housed in weatherproof enclosures, attached to the existing water-Cherenkov stations, as shown in Fig. 3. The scintillator light will be read out with wavelength-shifting fibers, which are bundled and attached to photomultiplier tubes Fig. 4. Since the surface stations are always working, the AugerPrime upgrade will acquire information for the full data set collected in the future [15].

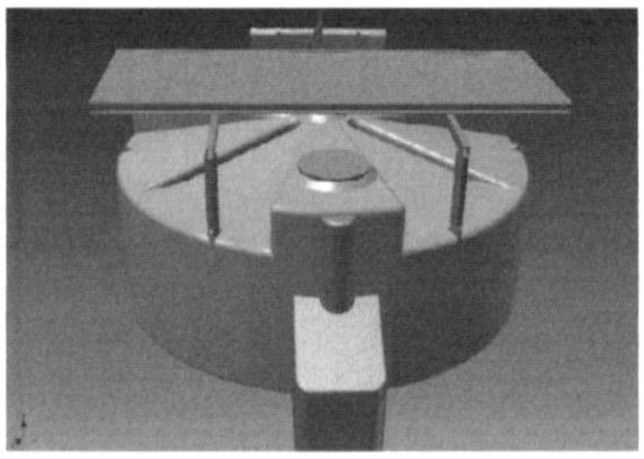

Figure 3. Drawing of SD station with scintillator detector (SSD) of area 4 m^2, housed in a weatherproof enclosure.

Figure 4. SSD detector with green wavelength-shifting fibers, which carry light to a photomultiplier tube(not shown).

The AugerPrime project includes other detector improvements. The dynamic range of the surface stations will be extended with the addition of a fourth photomultiplier. New electronics will be included, with faster sampling of the PMTs signals to a better identifications of the muon signals. New GPS will be implemented in each surface station to improve the timing accuracy and calibration. Also, some tests have been done with the fluorescence telescopes to extend measurements during nights with highly illuminated moon above the horizon. This will extend the duty cycle of the surface and fluorescence detector arrays of the Pierre Auger Observatory to 30% approximately. As a consequence, AugerPrime will have more precise measurements of the energy and X_{max} and therefore a better determination of the mass composition of the primary cosmic rays radiation.

In November 2015, the Auger Collaboration officially announced AugerPrime in Malargüe, Argentina, during a meeting with its International Finance Board and dignitaries from different collaborating countries. A renewed agreement was signed and it was establish that AugerPrime will be operate for 10 years. It is expected that, by 2018, the installation of the new detectors will be complete, in the meantime the Observatory will continue taking data.

References

[1] The Pierre Auger Coll., JCAP 1302, 026 (2013).
[2] The Pierre Auger Coll., Phys. Rev. D 90, 122006 (2014).
[3] F. Nerling et al, astro-ph/0506729 and M. Giller et al, J. Phys. G 31, 947-958 (2005).
[4] D. Gora et al, Astropart. Phys. **24**, 484-494 (2006).
[5] M. Giller et al, Int. J. Mod. Phys. A, **20**, 6821-6824 (2005).
[6] The Pierre Auger Coll., JCAP 08, 049 (2015). *Proc. of 33rd Ing. Cosmic Ray Conf., Rio de Janeiro* (2013).
[7] The Pierre Auger Coll., Phys. Rev. D 90 , 012012 (2014).
[8] M.Ave, R.Engel, J.Gonzalez, D.Heck, T.Pierog, and M.Roth, "Extensive Air Shower Universality of Ground Particle Distributions", *Proc. of 31st Int.Cosmic Ray Conf., Beijing* (2011).
[9] Hillas A.M., Acta Phys. Acad. Sci. Hung., 29 (1970), Suppl.3, 355 Hillas A.M. Proceedings of the 12 th International Conference on Cosmic Rays, vol.3, 1001, 1971.
[10] Piera L. Ghia, for the Pierre Auger Collaboration, "Statistical and systematic uncertainties in the event reconstruction and S(1000) determination by the Pierre Auger surface detector", *Proc. of 29st Int.Cosmic Ray Conf., Pune* (2005).
[11] Lafebre S. et al., Astrop.Phys.31 (2009) 243 (and refs there in).

IOP Conf. Series: Journal of Physics: Conf. Series **866** (2017) 012001 doi:10.1088/1742-6596/866/1/012001

[12] J.C.Espadanal, for the Pierre Auger Collaboration, "Measurement of the Muon Content of EAS with the Pierre Auger Observatory", *Proc of XXX-th IHEP 2014* (2014).
[13] The Pierre Auger Collaboration, "The Next Frontier in UHECR Research with an Upgraded Pierre Auger Observatory", *Proc of CSS2013, Mississippi* (2013).
[14] Ralph Engel, for the Pierre Auger Collaboration, "Test of hadronic interaction models with data from the Pierre Auger Observatory", *Proc. of 30th Int.Cosmic Ray Conf., Mexico* (2007).
[15] Gregory Snow, for the Pierre Auger Collaboration, "AugerPrime looks to the highest energies",CERN Courier, 56 (5): 29-31, June (2016).

IOP Conf. Series: Journal of Physics: Conf. Series **866** (2017) 012002 doi:10.1088/1742-6596/866/1/012002

Preliminary studies of the effect of thinning techniques over muon production profiles

G Tomishiyo and V Souza

Instituto de Física de São Carlos - USP - Av. Trabalhador São Carlense 400, Pq. Arnold Schimidt - São Carlos - SP - Brasil.

E-mail: tomishiyo@ifsc.usp.br

Abstract.
In the context of air shower simulations, thinning techniques are employed to reduce computational time and storage requirements. These techniques are tailored to preserve locally mean quantities during shower development, such as the average number of particles in a given atmosphere layer, and to not induce systematic shifts in shower observables, such as the depth of shower maximum. In this work we investigate thinning effects on the determination of the depth in which the shower has a maximum muon production ($X_{\rm max}^{\mu-{\rm sim}}$). We show preliminary results in which the thinning factor and maximum thinning weight might influence the determination of $X_{\rm max}^{\mu-{\rm sim}}$.

1. Introduction

Typical quantities of interest regarding the study of ultra-high energetic cosmic rays (UHECR) are their energy spectra, mass composition and distribution of arrival directions. Particle interaction properties are also studied in specific analysis. Regarding the mass composition and interaction properties studies, the most commonly used observable is the depth of shower maximum, X_{max} , defined as the atmospheric slant depth where the number of electromagnetic particles in the shower is maximum. The underlying theory that relates X_{max} to the primary particle composition and particle interaction properties was developed by Heitler in the 1940s [1, 2].

The Pierre Auger Collaboration makes direct measurements of X_{max} by using fluorescence detector (FD). The Collaboration analyzed eight years of data comprising more than two million events before cuts. The X_{max} evolution with energy shows a break at $10^{18.27}$ eV [3] and if air-shower simulations are used to interpret the data, it is possible to conclude the mass composition is getting heavier above this energy [4]. The Telescope Array Collaboration currently cannot discriminate between a proton dominated composition or a changing composition, due to low statistics limitations [5]. Moreover, the FD analysis suffers from a small duty cycle ($\sim 10-15\%$), since the telescopes can only be operated during clear moonless nights. Thus, one would greatly benefit from a second, independent observable for composition, exploiting Pierre Auger's surface detector (SD) stations which have a nearly 100% duty cycle.

Muons have been used as a important observable related to the primary particle mass and particle interaction properties [6, 7]. Muons are mostly generated by pion and kaon decay. The majority of pions and kaons are generated by the hadronic component of the shower, and since primary particles that differ in mass have different hadronic properties, it is natural to expect the muon content of the shower to be sensitive to the primary mass composition and to the hadronic interaction properties. Indeed, one would expect that, regarding muon production (the number of muons created at a given slant depth), heavy primaries should yield a shallower profile with a greater amplitude, since heavy primary particles generate muon rich showers and one can think of a primary particle of atomic mass A and energy E as A independent nuclei each with energy E/A (a reasonable approximation since the nuclear binding energies, in the MeV scale, are much smaller than the typical primary cosmic ray energy $> 10^{12}$ MeV).

In order to interpret the measurements of the observables, one needs to perform Monte Carlo simulations and compare the results to the measurements. Extensive air showers (EASs) are phenomena consisting of a large numbers of particles produced by cosmic rays that interacts with Earth's atmosphere. Since the total number of particles in an extensive air shower can be very large, a statistical technique - the thinning technique - is implemented in simulation software. However, by employing this technique there is, in principle, a risk of inducing some bias in the simulation results. So far, such biases were either not found or found to be much smaller than systematics due to experimental features [8, 9, 10, 11].

In this paper we present a preliminary study of the effect of the thinning on the depth of maximum muon production $X^{\mu-\mathrm{sim}}_{\mathrm{max}}$. The study is based only in the shower simulations, no detector simulaton is taken into account. The Pierre Auger Collaboration published results based on maximum muon production depth [6], however the analysis done in this paper are very different and much more sophisticated than the ones presented here. We will discuss the differences in section 3.1.

2. Thinning Algorithm

The thinning technique consists in following just a subset of particles in an EAS simulation: all the particles above a defined energy threshold ϵE_p (where ϵ is the thinning factor and E_p is the primary particle energy), and for particles below the threshold the technique adopts different policies depending on the interaction that gave them birth. If all particles in the interaction have an energy below threshold, then the algorithm will follow a single particle with probability e/E_{sum}, where e is its energy and E_{sum} is the sum of all particle energies. To this particle, the weight $E_{sum}/e * w$ is assigned to statistically preserve shower quantities (w is the weight of the mother particle; unthinned particles have unit weight). If, however, some of the particles have energies greater than the thinning threshold, one will follow all those particles above threshold, and follow a sub-threshold particle of energy e with probability e/E_t (where E_t is the sum of all particle energies below the thinning threshold). If a particle below the thinning threshold is followed, a weight $E_t/e * w$ is assigned to it.

The above technique is know as the Hillas thinning [8]. Alternative thinning techniques, such as the statistical thinning described in [12] are also employed, depending on what observables one is interested in analyzing with the simulation software. Additionally, statistical fluctuations are considerably reduced by setting a maximum particle weight: a particle with weight $w > w_{max}$ is no longer subject to thinning. The thinning threshold and maximum weight sometimes differ for hadronic and electromagnetic particles, since the former are less numerous than the latter. Finally, a radial thinning technique can be employed, which consists in defining a radial scale r_{max}, and then retaining all particles with radial position (with respect to the shower axis) $r > r_{max}$ and otherwise selecting the particle with probability r/r_{max} and multiplying its weight by r_{max}/r.

Thinning techniques are known to preserve mean shower quantities [8], to not induce bias in observables such as the depth of shower maximum [9, 11] or the risetime [10]. In other cases, it has been shown that the spurious fluctuations introduced by the thinning in the risetime are smaller than statistical fluctuations due to detection and reconstruction [10]. The thinning

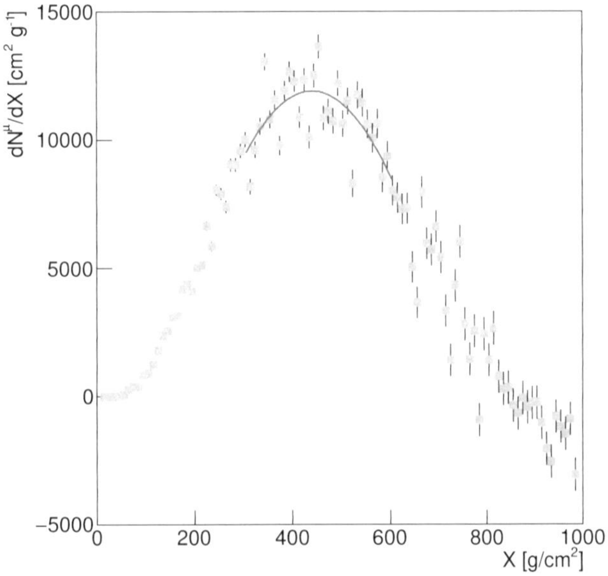

Figure 1. Example of the evolution of the number of muons as a function of depth. All muons produced in the shower development were taken into account. Shower simulated with `CORSIKA` and `QGSJETII-04` and `UrQMD 1.3cr` hadronic interaction models at 65° zenith angle. Primary particle was a carbon nucleus with energy 1×10^9 GeV. Red line shows a fit of a second order polynomial to the points used to calculate the maximum of the profile $X^{\mu-\mathrm{sim}}_{\mathrm{max}}$. The thinning factor was set to 10^{-6} and $w_{max} = 10^4$.

technique greatly improves computational speed and reduces required storage space. Without it, it would not be possible to simulate large samples of UHECR showers.

3. Muon production profile at its maximum

We studied the muon production profiles and its maximum using Monte Carlo simulations of air showers. The simulation software of choice was `CORSIKA 7.5000` [13]; we used `QGSJetII-04` [14] as the high energy interaction model and `UrQMD 1.3cr` [15] as the low energy model. The primary particle was fixed to 1×10^9 GeV with zenith angle of 65°. The geomagnetic field was chosen so as to be the same as in Malargue, in Mendoza, Argentina (the site of the Pierre Auger Observatory).

In order to investigate thinning effects on muon production profiles we simulated showers with two thinning factors (10^{-6} and 10^{-5}) and three maximum weight for the hadronic and electromagnetic thinning $w_{max} = 10^3$, 10^4 and 10^5 . In our study, we consider all muons produced in the shower development. The number of muons in the shower are sampled in steps of 5 g/cm^2 along the slant depth. We used the longitudinal tables produced in the output of CORSIKA and subtracted the number of muons from consecutive steps. But this procedure we evaluate the "production rate" of muons, because in each step muons are created and decay and we can not separe the two effects in our estimation.

Figure 1 shows an example of the development of the muon profile. The simulated muon profile was fitted by a second order polynominal as shown by the red line in the figure. We used

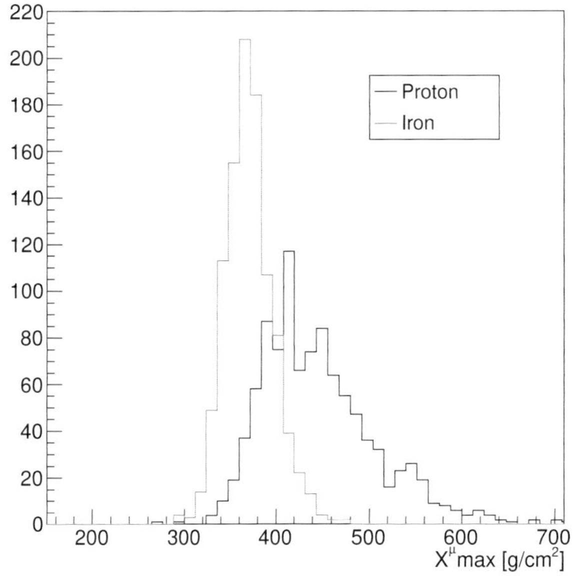

Figure 2. Simulated X^{μ}_{max} distribution for proton and iron nuclei as primary particles. All muons produced in the shower development were taken into account. Showers simulated with `CORSIKA` and `QGSJETII-04` and `UrQMD 1.3cr` hadronic interaction models at 65° zenith angle. Primary particles had energy 1×10^{9} GeV. The thinning factors were set to 10^{-6} and $w_{max} = 10^{4}$.

the fitted function to calculate $X^{\mu-\mathrm{sim}}_{\mathrm{max}}$. Figure 2 shows the calculated $X^{\mu-\mathrm{sim}}_{\mathrm{max}}$ for all proton and iron nuclei primaries simulated. As expected, iron primaries generate narrow profiles with a shallow maximum as compared to proton primaries.

3.1. Differences from the Pierre Auger Collaboration analysis

The Pierre Auger Collaboration has elaborated a sophisticated data analysis to extract from the measured air showers the relevant information from the muon profile. References [6, 16] summarizes the data analysis procedure. The data analysis starts with the arrival time of the particles in the SD stations. A detailed selection procedure was applied in order to minimize systematic and statistic uncertainties, maximize the signal-to-noise ratio and deconvolve detector limitations. Only stations far from the shower axis ($r > 1700$ m) are used and only events with zenith angle within the interval [55°, 65°] and energy between log(E/eV) = 19.3 − 20 are considered.

The arrival time of particles is converted into a muon production depth using a model described in refences [17, 18, 19]. Note that by this procedure, only muons arriving on ground contributes to the measured muon profile, therefore low energy muons which might decay in the shower development are not taken into account. A Gaisser-Hillas function is fitted to the longitudinal muon production depth profile and X^{μ}_{max} is determined. Finally, in an internal publication [20], the Pierre Auger Collaboration has evaluated the effect of thinning factors in the muon production depth analysis and it has shown to be negligible when thining levels below 10^{-6} are used.

The summary of the data analysis of the Pierre Auger Collaboration presented above clarifies the difference between the preliminaries studies presented here and the detailed analysis done by the Pierre Auger Collaboration. We call the main quantity studied here $X_{\rm max}^{\mu-\rm sim}$ to make it explicit that we take all muons produced in the simulation (sim) irrespectevely of their distances from the shower core. Our study also considers all muons produced in the shower independently if they would hit ground or not. Moreover we made our study for one fixed energy 1×10^9 GeV and zenith angle 65°.

Finally, we would like to make it clear that the quantity calculated here $X_{\rm max}^{\mu-\rm sim}$ can not be directely compared to the quantity used by the Pierre Auger Collaboration in its publications and therefore the conclusion can not be extrapolated.

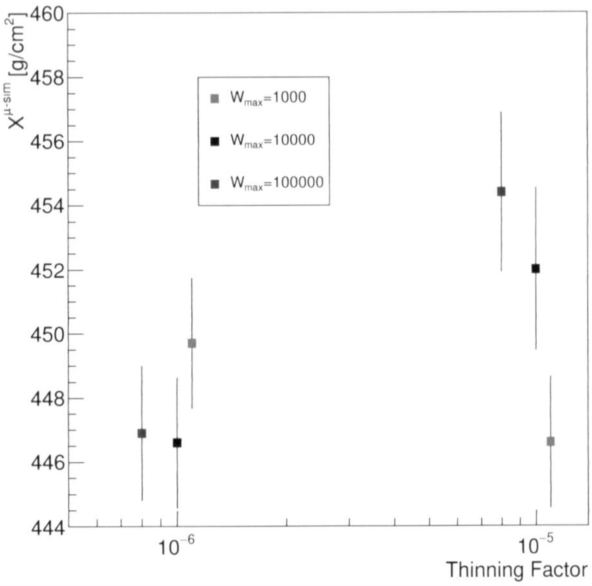

Figure 3. $X_{\rm max}^{\mu-\rm sim}$ for different thinning factors and weight limits. All muons produced in the shower development were taken into account. Shower simulated with `CORSIKA` and `QGSJETII-04 UrQMD 1.3cr` hadronic interaction models at 65° zenith angle. Primary particle was a proton with energy 1×10^9 GeV. Points were slightly displaced horizontally so that the error bars would not overlap.

4. Results

Figure 3 displays the preliminary results obtained. Each point in the figure corresponds to the $X_{\rm max}^{\mu-\rm sim}$ from 1000 proton simulated showers, for a given thinning factor and weight limit. Figure 3 suggests that the thinning factor and the maximum weight limit might have an effect on the $X_{\rm max}^{\mu-\rm sim}$ values. The effect is not trivial and the error bars are too large for a definitive conclusion. Currently, we are investigating this matter by writing a toy Monte Carlo code, which implements the same interaction and decay procedure as `CORSIKA` and running other thinning factors and weight limits.

5. Conclusion

The muon content of an EAS provides valuable information, which can be used to study the primary particle and to discriminate between competing interaction models. For this reason it is crucial that all possible errors be kept in check. In this work we analysed the effect of the thinning factor and maximum weight on the $X_{\max}^{\mu-\text{sim}}$ parameters. It has been shown that $X_{\max}^{\mu-\text{sim}}$ can be affect by different values of thinning factor and maximum weight. Further investigations are under way.

Acknowledgments

We are grateful to Conacyt for support through the Mesoamerican Center for Theoretical Physics and to Coordenação de Aperfeiçoamento de Pessoal de Nível Superior (CAPES - Brasil) for providing a scholarship for this research project. The authors thank FAPESP (2014/19946-4) for the support. VdS thanks CNPq.

References

[1] Heitler W 1947 *The Quantum Theory of Radiation* (Oxford University Press)
[2] Matthews J 2005 *Astroparticle Physics* **22** 387–97
[3] The Pierre Auger Collaboration 2014 *Phys. Rev. D* **90**(12) 122005–30
[4] The Pierre Auger Collaboration 2014 *Phys. Rev. D* **90**(12) 122006–18
[5] Barcikowski, E and Bellido, J and Belz, J and Egorov, Y and Knurenko S and Souza, V et al 2013 *EPJ Web Conf.* **53** 1–14
[6] Aab A and et al 2014 *Phys. Rev. D* **90** 012012–27
[7] Unger M 2016 *Proc. of Int. Symp. for Ultra-High Energy Cosmic Rays 2014* **010020** 1–7
[8] Hillas A M 1997 *Nuclear Physics B* **B** 29–42
[9] Gorbunov D S, Rubtsov G I and Troitsky S V 2007 *Phys. Rev. D* **76** 043004–11 (*Preprint* 0703546)
[10] Bruijn R, Schmidt F, Ilee J and Knapp J 2009 *Proc. of the 31st Int. Cosmic Ray Conf.* **2** 1–4
[11] Todero C J, Souza V and Bellido J A 2013 *Astroparticle Physics* **47** 18–30
[12] Kobal M 2001 *Astroparticle Physics* **15** 259–73
[13] Heck, D et al 1988 *CORSIKA: A Monte Carlo Code to Simulate Extensive Air Showers* FZKA–6019 (FZKA)
[14] Ostapchenko S 2011 *Phys. Rev. D* **83** 1010–869
[15] Bleicher, M et al 1999 *J.Phys.G:Nucl.Part.Phys.* **25** 1859–96
[16] Laura Collica for the Pierre Auger Collaboration 2016 *Eur. Phys. J. Plus* **131**
[17] Cazon L, Vázquez R and Zas E 2005 *Astroparticle Physics* **23** 393–409 ISSN 0927-6505
[18] Cazon L, Vázquez R Watson A and Zas E 2004 *Astroparticle Physics* **21** 71–86 ISSN 0927-6505
[19] Cazon L, Conceição R, Pimenta M and Santos E 2012 *Astroparticle Physics* **36** 211–223 ISSN 09276505
[20] Bueno A, Garcia-Gamez D and Bueno L M *GAP-Notes - Pierre Auger Internal Notes* **2015-34**

IOP Conf. Series: Journal of Physics: Conf. Series **866** (2017) 012003 doi:10.1088/1742-6596/866/1/012003

Study of Risetime as a function of the distance to the Shower Core in the Surface Detector (SD) of the Pierre Auger Observatory

Hernán Castellanos Valdés [a], **Karen Salomé Caballero Mora** [a,b]

[a]Universidad Autónoma de Chiapas; [b]Mesoamerican Centre for Theoretical Physics (MCTP)

E-mail: hernancastellanosvaldes@gmail.com

Abstract. Cosmic Rays (**CR**) are high energy particles which come from the universe. When one of those particles enters to the earth's atmosphere it produces an air shower, conformed by secondary particles in which the initial energy is distributed. The Pierre Auger Observatory, located in Argentina, is dedicated to the study of those events. One of the main goals is to find out where those **CR** are coming from and which kind of chemical composition do they have. In this work we show the status of a study of the risetime ($t_{1/2}$) as a function of the distance to the shower core (near to the air shower's axis) for different zenith angles and energies, obtaining a new variable that will be compared with other variables used by the Observatory. The main objective of this study is to better understand risetime as a mass composition sensitive parameter of **CR**.

1. Introduction

Ultra High Energy Cosmic Rays (UHECR) are high energy particles that come from the universe. Due to their low frequency they can only be studied indirectly when one of these particles enters to the Earth's atmosphere and it produces an air shower, conformed by secondary particles in which the initial energy is distributed [1].

1.1. Pierre Auger Observatory

The Pierre Auger Observatory, located in Argentina, is dedicated to the study of those events. One of the main goals is to find out where those UHECR are coming from and which kind of chemical composition do they have. The Observatory has two kind of detectors, the Surface Detector (SD), and 24 Fluorescence Detectors (FD) [2]. The Observatory is also equipped with detectors for atmospheric monitoring and hosts a radio array for the study of EAS at MHz frequencies among other upgrades recently deployed and in process to be deployed [3].

1.2. Surface Detector

The Surface Detector covers an area of 3000 km^2 with 1660 water Cherenkov detectors , with 12000 l ultrapurified water and three Photomultiplier Tubes each one.

2. Offline

The Offline [4] software framework has been developed within the Pierre Auger Collaboration to perform the computing tasks. It is designed to provide all the functionalities to process data from the shower detector, including simulation and reconstruction, and it is implemented in C++. The Offline framework provides an easy-to-use interface, making possible to modify the codes and to include new tasks, completely hiding the detailed internal mechanisms of how and where the data are taken from. There is also a general purpose from SD, FD as well as hybrid event reconstruction. For this work the *advanced data summary trees* ADST data samples produced with Offline are used, to manipulate them only a valid ROOT [5] installation is necessary. Figure 1 shows an event observed by the FD using the Offline graphic tools.

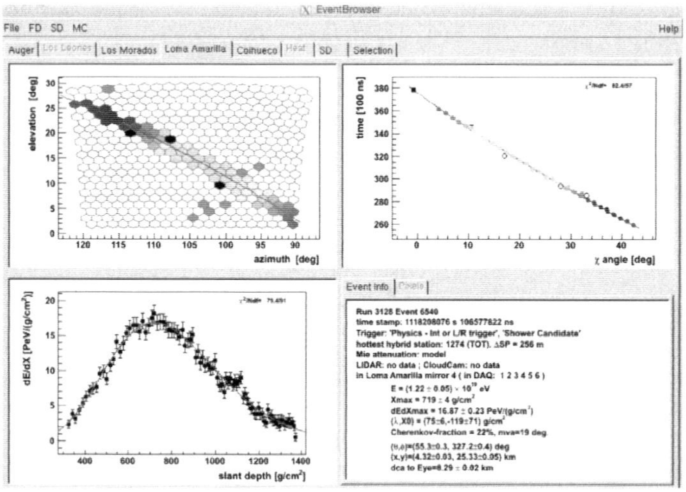

Figure 1. An event measured by the FD in the viewer of Offline.

3. Risetime

The $t_{\frac{1}{2}}$ is the time it takes to the integrated signal recorded by the SD to rise from 10% to 50% of the final value, a diagram of the risetime for a simulated signal is shown in Figure 2 [6]. It is a mass composition sensitive parameter of UHECR since for different primary particles there is different risetime value. Figure 3 [7] shows that the risetime for iron particles is smaller than that for light particles as proton at a distance of 1000 m from the shower core, from simulated showers. The signal produced by different primaries results in different risetimes due, among other detector effects, to the depth of the first interaction in the atmosphere as shown in Figure 4 [6]. For heavy primaries the interaction occurs early, then the electromagnetic component of the air shower attenuates at ground more than it happens when the interactions occurs deep in the atmosphere as in the case of light particles. Therefore, for the first case the risetime is smaller than for the second one.

4. Work Description

Typically a value of $t_{\frac{1}{2}}$ is considered for each event, this is the one a detector located at 1000 m from the shower core would have, no matter the energy or zenith angle of the event. Nevertheless this value shows a large spread [6]. This does not allow to directly use this parameter as mass composition sensitive. The main objective of this study is to find a distance to the shower core, zenith angle or energy dependent, for any particular event which might not necessarily be 1000 m. It is expected this parameter presents a smaller spread than the one of $t_{\frac{1}{2}}$ at 1000 m. Once the distance as function of the zenith angle or of the energy is found, the corresponding $t_{\frac{1}{2}}$ will

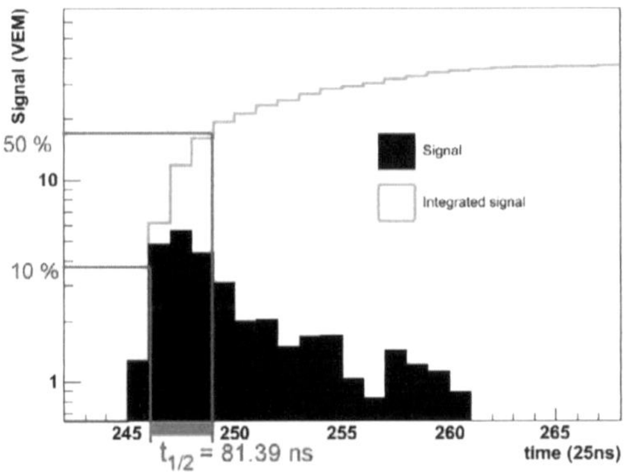

Figure 2. Diagram of the definition of $t_{\frac{1}{2}}$ [6].

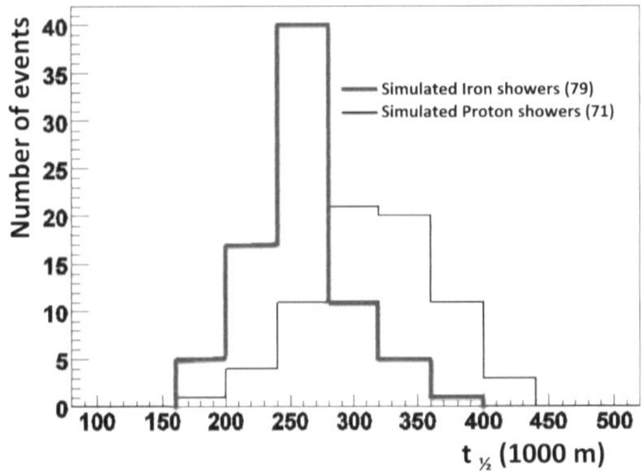

Figure 3. Comparison of $t_{\frac{1}{2}}$ for heavy and light particles, from simulated showers [7].

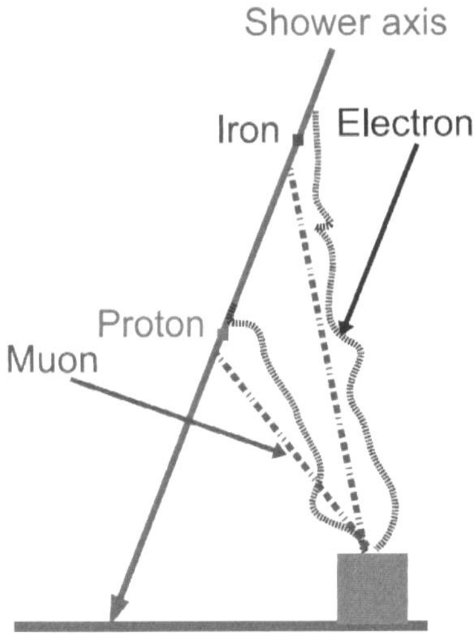

Figure 4. Diagram of the difference between the interaction of heavy particles and protons with the atmosphere [6].

be tested as mass composition sensitive parameter to be compared with other SD parameters. The first attempt is to find the distance as a function of zenith angle for a similar energy. Events with angles around 30°- 39°, 40°- 49°and 50°- 59°are analysed. Fits of the $t_{\frac{1}{2}}$ as a function of shower core for a certain event are performed. The forms of the used fts are the following:

$$f(x) = 40 + ax + bx^2 \text{ [6] and [8]}$$
$$f(x) = 10 + \sqrt{a^2 + bx^2} - a \text{ [9]}$$
$$f(x) = 40 + ax^b$$

Zenith Angle (∘)	Energy x 10^{19} (eV)	InD (m)
36.56	2.29	765
44.77	1.37	910
57.57	1.62	1095

Table 1. Examples of characteristic distances for $t_{1/2}$ for different zenith angles and similar energies.

In a range of 600 m to 1500 m of distance to the shower core. The intersection of the fits for the same event will be the distance we want to find. This first study shows that there is a different intersection distance of the fitted functions of $t_{\frac{1}{2}}$, for events with different zenith angles. That distance might be used for performing more accurate studies on mass composition. A total of 15 events have been analysed showing similar results. Table 1 shows the preliminary results of three of the analysed events. This common point is found by eye. Currently a method to find the intersection point through a code for several events and setting better quality cuts is being performed. The dependence of the distance with the energy will be also explored and after finding a good characterization of such intersection distance, studies on mass composition will be performed.

5. Conclusion

The $t_{\frac{1}{2}}$, as a mass composition sensitive parameter, could be optimized if the characteristic length for each event is found. the results about composition could improve with respect to the data taken at 1000 m.

6. Aknowledgements

- To CONACyT for the support granted through the project CB 243290
- To L'ORÉAL México, CONACyT, UNESCO, CONALMEX and AMC for the support granted through the "Beca para Mujeres en la Ciencia 2014" for Karen Salomé Caballero Mora.

References

[1] N.Nagano and A. A. Watson, Rev. Mod. Phys. 72, 689(2000), *Observations and implications of the ultrahigh-energy cosmic rays.*
[2] The Pierre Auger Collaboration, NIM A 798 (2015) 172-213, *The Pierre Auger Cosmic Ray Observatory*
[3] The Pierre Auger Collaboration, arXiv:1604.03637v1, astroph-IM, The Pierre Auger Observatory Upgrade Preliminary Design Report (2016).
[4] S. Argiro. et al, Nucl. Instrum. Meth. A580, 1485 (2007).
[5] *https://root.cern.ch/*
[6] K.S Caballero-Mora *Composition studies of Ultra High Energy Cosmic Rays using Data of Pierre Auger Observatory*, Ph.D. Thesis, 2010, KIT, Germany
[7] K.S Caballero Mora, et al *Composition studies using the surface detector of the Pierre Auger Observatory*, J.Phys.Conf.Ser.116:012007, 2008
[8] B. Smith, *The Mass Composition of Cosmic Rays Above 1 EeV Inferred Using the Spread in Arrival Times of Air Shower Particles*, 2008, PhD thesis, Department of Physics and Astronomy, University of Leeds, UK.
[9] C. Jarne *Estudios de composición primaria en base a distribuciones temporales en el detector de superficie del Observatorio Pierre Auger*, 2015, PhD thesis, Universidad Nacional de La Plata, IFLP.

IOP Conf. Series: Journal of Physics: Conf. Series **866** (2017) 012004 doi:10.1088/1742-6596/866/1/012004

Analysis of Solar Events in 2015 with the HAWC Gamma Ray Observatory

E. Tirado[1], E. Moreno[1] and A. Lara[2] for the HAWC Collaboration

[1] Meritorious University Autonomous of Puebla
[2] Institute of Geophysics UNAM, Department of Space Sciences

E-mail: etb.tirado@gmail.com, emoreno.emb@gmail.com and alara.unam@gmail.com

Abstract. The work consists in finding and analyse solar events, such as the solar flares and coronal mass ejections and study the variation in the data counting, which can be observed by the high - rise light Cherenkov in water "HAWC" observatory, consist of 300 tanks of ultrapure water, where each of them has 4 photomultiplier tubes. HAWC is dedicated to the observation of gamma and cosmic rays with energies of TeV and has an opening that covers more of 15 % of the sky. With its wide field of view, the observatory is exposed two - thirds of the sky during each 24 hours cycle. The detector has to main data acquisition systems: the particular event and the total sum or Scalers, in this work is will make use of the counting system called TDC Scaler, which records the count rate of the number of particles which impacted on the detector in a time range.

1. Introduction

The High Altitude Water Cherenkov (HAWC) observatory is an air-shower array composed of 300 water Cherenkov detectors (WCDs) currently totally constructed within the Pico de Orizaba National Park (México) by a Mexican-United States collaboration. The main purpose of HAWC is to observe gamma rays and cosmic rays between 100 GeV and 100 TeV, TeV gamma rays are the highest energy photons ever observed.

Milagro was the first generation experiment that operated at a lower altitude of 2630 meters close to the Los Alamos, New Mexico. The pool design of the Milagro experiment has been replaced in the HAWC experiment by a modular design (array of WCDs). The WCD is designed to detect the Cherenkov light produced by secondary particles using photomultipliers tubes (PMTs) immersed in water.

[1] Avenida San Claudio y 18 Sur, Colonia San Manuel, Edificio FM5-206, Ciudad Universitaria, C.P. 72570, Puebla, México.
[2] Instituto de Geofísica, Circuito de la Investigación Científica s/n, Ciudad Universitaria, Del. Coyoacán, C.P. 04150, Ciudad de México.

The Sun is the closest star to Earth, and because of this is the only star that can be seen with spatial, temporal and spectral resolution without comparison in Astronomy. For this reason the sun is the laboratory base to study and understand the physical phenomena occurring in other stars of similar characteristics.

The Sun is also an active star in whose atmosphere occur violent and energetics events, called solar flares and coronal mass ejections. During this events the Sun emits radiation at very high energies, the particles are accelerated, and a huge amount of material is ejected to the interplanerary medium.

These remainder of the solar activity reaches the Earth and other planets, their atmospheres and magnestosphere. The origin of this active phenomena is closely related to the existence of localized areas of intense magnetic fields referred to as active regions.

For this reason in this work we study the variation of the scaler counting during solar events regitered in 2015, and specially during a Forbush Decrease (FD) event generally indicates a decrease in the cosmic-ray count rate caused by transient interplanetary events which are strictly related to a solar mass ejection activity (Cane 2000).

2. Methodology

To observe the effects of solar modulation of galactic cosmic rays, we revised the PMTs scaler rates considering the effect of pressure variabilty. The correction used is described by the following equation

$$R_{corr} = R_{meas} - Rp + \langle R_{meas} \rangle \qquad (1)$$

were R_{corr} is the corrected scaler rate, R_{meas} is the single PMT measured rate and $Rp = a*P + b$ is the linear function used to describe the relation between the scaler rate and the measured atmospheric pressure (P).

For the realization of this work it will be used the software called Python, together with specialized library for solar issues called Sunpy, which helps to acces to the public data of the GOES Observatory of the NASA (NationalAeronautics and Space Administration). With these data, we can know which of the solar events registered in 2015 are interesting for studying in the HAWC Observatory.

Table 1. Solar flares of great intensity in 2015.

Date	Flare Class	Max, UT
03-03-15	M8.2	00:27:00
07-03-15	M9.2	22:22:00
11-03-15	X2.1	16:22:00
05-05-15	X2.7	22:11:00
22-06-15	M6.5	18:23:00

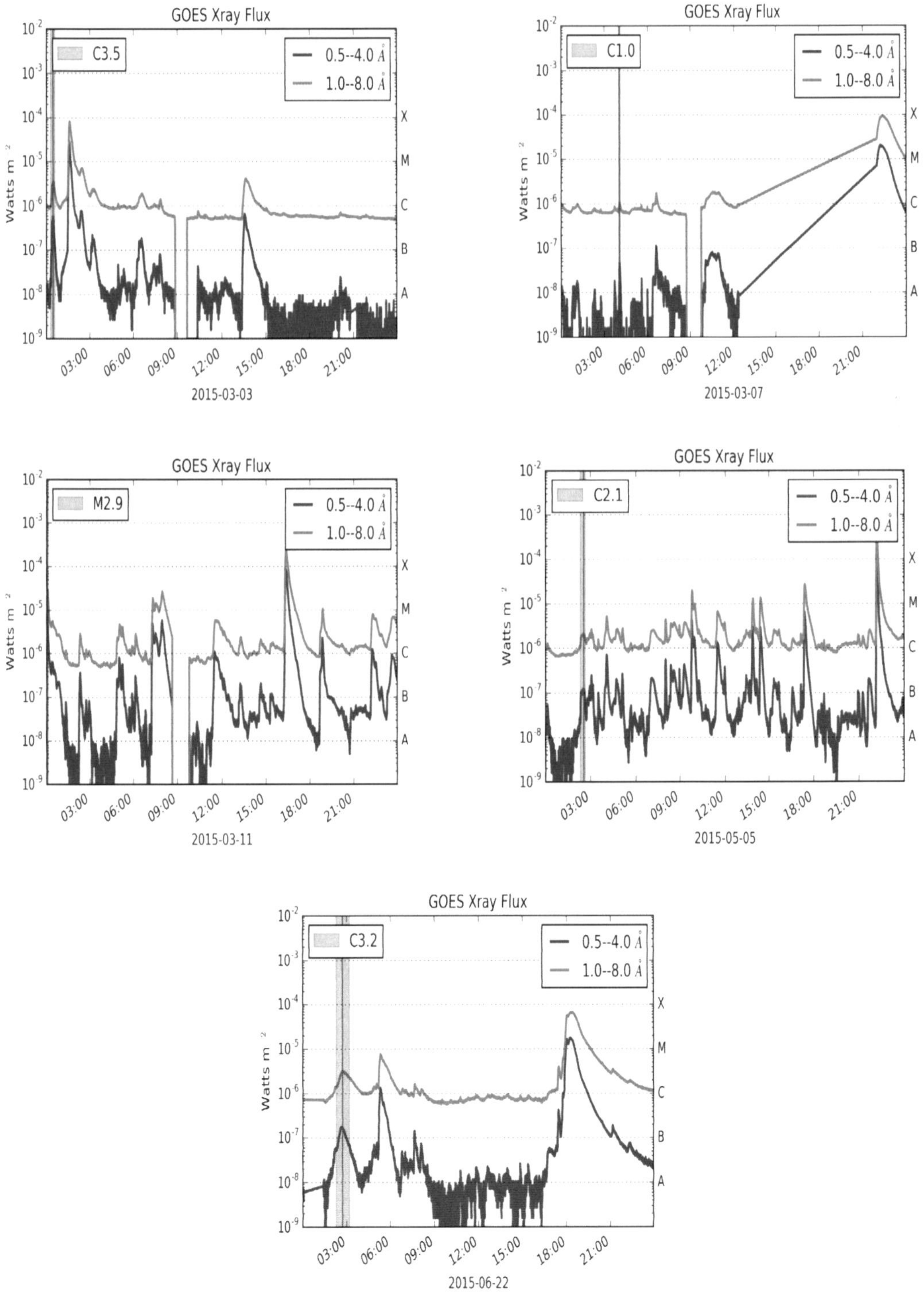

Figure 1: In this figure we show the plots of emission of X Ray flux for each of the solar events shown in Table 1. These graphics are made with public data from the Observatory GOES 15.

3. Discussion and Results

For the analysis of the HAWC observatory data (TDC Scaler). We desing a program in Python, which processes data from the PMTs that are generated every 10 minutes and contain around 60000 measurements for each PMT, and then plot the information obtained to verify the Variation of the scaler counting during a solar event, for example in solar flares and coronal mass ejections are enhancements in the counts for few minutes to hours and in the case of interplanetary coronal mass ejections are decreases in the counts for few hours to days.

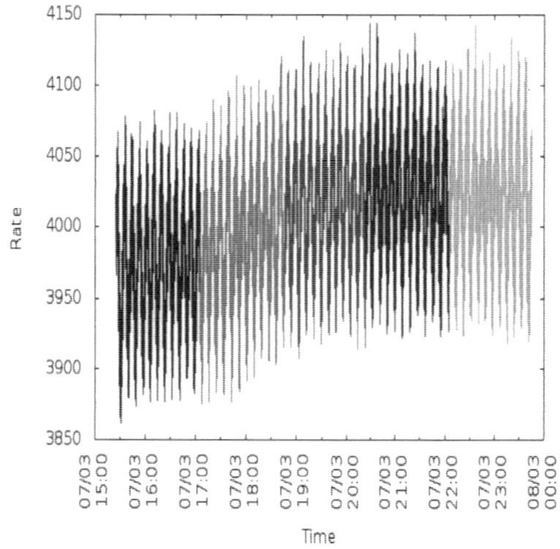

Figure 2a: Plot that show Rate vs. Time of the 07-03-15 solar event.

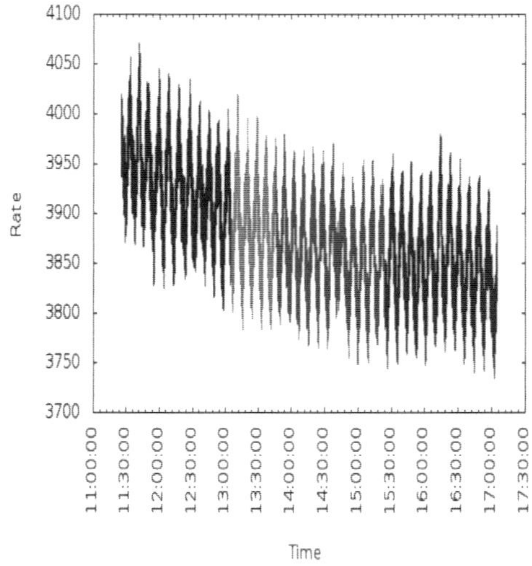

Figure 2b: Plot that show Rate vs. Time of the 11-03-15 solar event.

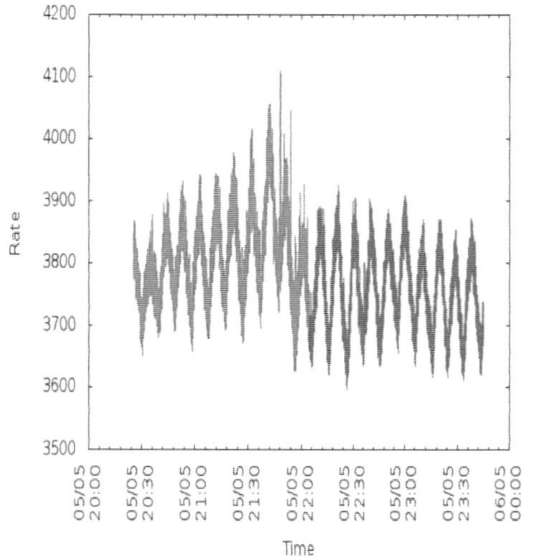

Figure 2c: Plot that show Rate vs. Time of the 05-05-15 solar event.

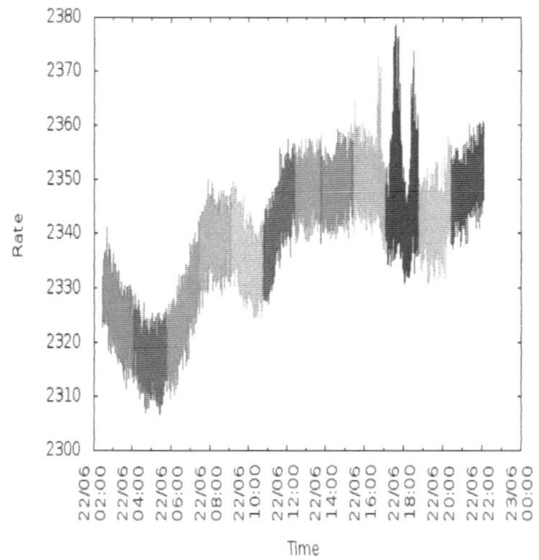

Figure 2d: Plot that show Rate vs. Time of the 07-03-15 solar event.

In Figure 2 we can observe that in the subfigure 2b and 2c we have an enhacemnet of the counting data, which that implies that the solar event we are talking about is solar flare and in figure 2d we can observe that the solar events is an interplanetary coronal mass ejections.

4. Summary and Conclusions

In this work we analyze the HAWC observatory TDC Scaler data to find long transient solar events (solar flares and coronal mass ejections). We studied with the help of the plots obtained for solar events of the days listed above in Table 1 and we can study the variability of the counting data reported by Cane in 2000 with the HAWC Observatory.

5. Aknowledgments

This research has made use of SunPy, an open-source and free community-developed solar data analysis package written in Python.

References

A. Abdo, R. Atkins, W. Benbow, and et al. Recent Results from the Milagro Gamma Ray Observatory. Nuclear Physics B Proceedings Supplements, 151:101–107, Jan 2006. doi: 10.1016/j.nuclphysbps.2005.07.018.

U. Abeysekara, J. A. Aguilar, S. Aguilar, R. Alfaro, and et al. On the sensitivity of the HAWC observatory to gamma-ray bursts.Astroparticle Physics, 35:641–650, may 2012. doi: 10.1016/j.astropartphys.2012.02.001.

U. Abeysekara, R. Alfaro, C. Alvarez, and et al. The HAWC Gamma-Ray Observatory: Design, Calibration, and Operation. ArXiv e-prints, sep 2013a.

U. Abeysekara, R. Alfaro, C. Alvarez, and et al. Sensitivity of the High Altitude Water Cherenkov Detector to Sources of Multi-TeV Gamma Rays. Astroparticle Physics, 50:26–32, Dec 2013b. doi: 10.1016/j.astropartphys.2013.08.002.

R. Atkins, W. Benbow, D. Berley, and et al. Milagrito, a TeV air-shower array. Nuclear Instruments and Methods in Physics Research Section A, 449(3): 478 – 499, 2000. ISSN 0168-9002. doi: 10.1016/S0168-9002(00) 00146-7. URL http://www.sciencedirect.com/science/article/pii/S0 168900200001467

J. W. Bieber, J. M. Clem, M. L. Evenson, and et al. Latitude survey observations of neutron monitor multiplicity. Journal of Geophysical Research: Space Physics, 109(A12):2156–2202, 2004. doi: 10.1029/2004JA010493. URL http://dx.doi.org/10.1029/2004JA010493

H. V. Cane. Coronal Mass Ejections and Forbush Decreases. Space Science Reviews, 93: 55–77, Jul 2000. doi: 10.1023/A:1026532125747.

E. Carrasco, A Carramiñana, R. Avila, and et al. Weather at Sierra Negra: 7.3-yr statistics and a new method to estimate the temporal fraction of cloud cover, 398:407–421, sep 2009. doi: 10.1111/j.1365-2966.2009.15163.x.

IOP Conf. Series: Journal of Physics: Conf. Series **866** (2017) 012005 doi:10.1088/1742-6596/866/1/012005

Characterization of muon simulations for the scintillator detector BATATA located at the Pierre Auger Observatory

Itzel Amayrani Martínez Salazar[a], Dr. Gustavo Medina Tanco[b], Dr. Karen Salomé Caballero Mora[a], for the Pierre Auger Collaboration[c]

[a]Facultad de Ciencias en Física y Matemáticas, UNACH; [b]Institute Nuclear Sciences, UNAM; [c] Observatorio Pierre Auger, Av. San Martin Norte 304, 5613 Malargüe, Argentina (Full author list: *http : //www.auger.org/archive/authors_2015_12.html*)

E-mail: itzel.msalazar@gmail.com

Abstract. Ultra-High energy Cosmic rays (UHECR) reach us from our Galaxy and from extragalactic space. Their origin is still unknown and their study is one of the main aims of the Pierre Auger Observatory. The underground detector Buried Array Telescope at Auger (BATATA) was designed as part of the low energy enhancement of the Observatory. This telescope is a 3 m side cube with the main goal of analyzing the muon component from air showers produced by primary Cosmic Rays (CR) and quantifying the electromagnetic contamination of the muon signal as a function of depth in the 10^{17} and 10^{18} eV energy range. The objective of this work is to characterize the functionality of simulations developed exclusively for BATATA in order to understand the behavior of the muon component in the detector. This is done by comparing the simulations with analytical calculations. In this work, results on comparisons of the muon component are presented.

1. Introduction

The Pierre Auger Observatory is located in the southern hemisphere in the province of Mendoza, Argentina with an area of 3000 km^2. It uses two types of detectors, fluorescence telescopes and water-Cherenkov stations, therefore it is a hybrid detector, which aims to determine the nature, energy and origin of cosmic rays with energies over 10^{17} eV (See Figure 1). The Surface Detector (SD) consists of an array of 1660 water-Cherenkov detectors separated by 1.5 km, each station is independent with 3.6 m in diameter, 1.2 m depth and a capacity of 12 tons of pure water. The Observatory has 24 Fluorescence Detectors (FD) distributed in 4 buildings. These detectors operate at night when there is little background light so that they only detect the light emitted by the cascade in the atmosphere [1]. In recent years, the Pierre Auger Collaboration decided to expand the detection range to observe cosmic rays of lower energy (up to 10^{17} eV). The enhancements include fluorescence telescopes at high elevation (HEAT), underground muon detectors (AMIGA and BATATA) and detectors for radio (AERA) and microwave emissions of atmospheric showers. As part of the upgrade called AugerPrime, additional muon detectors are being depoyed in the original SD [2].

IOP Conf. Series: Journal of Physics: Conf. Series **866** (2017) 012005 doi:10.1088/1742-6596/866/1/012005

1.1. Description of the detector BATATA

Buried Array Telescope at Auger (BATATA) is an additional prototype for the AMIGA counter [3]. It was designed as part of the low energy enhancement of the Observatory, with the main aim of analysing the muon component from air showers produced by primary CRs and quantifying the electromagnetic contamination of the muon signal as a function of depth in the 10^{17} and 10^{18} eV energy range. This telescope is a 3 m side cube installed underground. It consists of six layers of plastic scintillator, buried at a depth between 30 cm and 2.5 m (See Figure 2). Each layer in a plane is 4 m^2 and composed by 50 rectangular bars of 4 cm x 2m, oriented at a right angle whith respect to its companion layer, which gives an xy-coincidence of 4x4 cm^2 [3] and [4]. BATATA is an independent detector, which can be used to perform studies of the interactions taking place in the air shower through the muonic component.

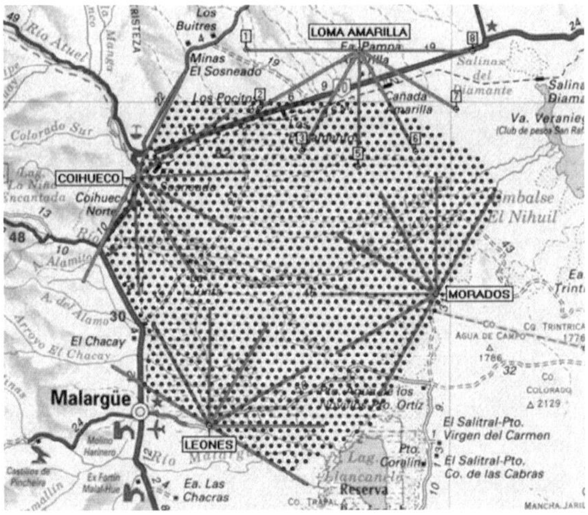

Figure 1. Area of the Pierre Auger Observatory in Mendoza, Argentina. Blue dots are SD stations. Four FD sites yellow over the observatory.

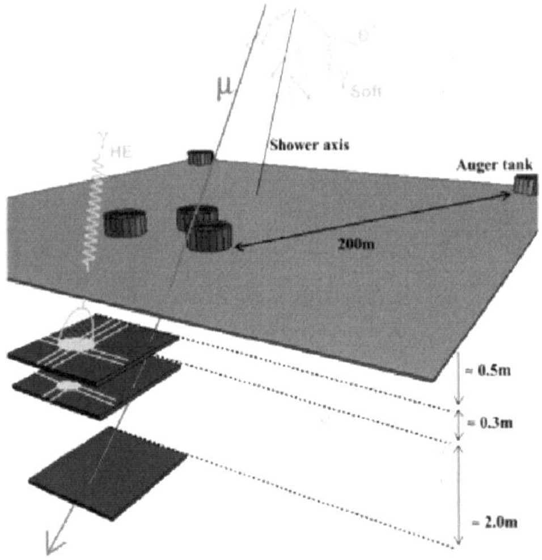

Figure 2. BATATA structure (see text).

2. Simulations

A code exclusive for BATATA has been developed at UNAM [5], based on python. This code, called in this document *cascade*, aims to describe the detection probability of muons and electrons at the detector based on analytical calculations taking into account the propagation of such particles through BATATA. It is important to mention that another simulation code has also been developed for BATATA, based on the Geant4 tool, to monitor the soil particles in *Malargüe* observatory region, see [3] and [4], this is not used in this work.

2.1. Analytical calculation of the detection probability

Analytical calculations were developed by Dr. Gustavo Medina Tanco from the Institute Nuclear Sciences, UNAM (private communication). Those calculations describe the detection probability of muon and electromagnetic components, based on the propagation of such particles through the detector. These calculations can be obtained for a given particle with different initial conditions using the software **Mathcad** (version 14.0). For more information on these calculations please see Appendix A in [6].

2.2. Description of "cascade" code

The code provides a simulation of muonic and electronic components from low energy particle interactions in the atmosphere. It is possible to throw muons and electrons of different energies and different geometries through the detector for pattern recognition based on statistics. The spread of particles arriving to the detector is represented, in the simulation, by the intersection of a cylinder with the layers which form the detector. That intersection has the shape of an ellipse as shown in Figure 3.

The *cascade* code is handled with two general scripts: **escritura.py** and **mezcla**, which work as follows:

escritura.py: is a code for writing data configuration parameters.

mezcla: is the master script, it runs once you have configured the escritura.py code from a terminal such as: ./mezcla.

3. Description of the method

The goal is to verify that the *cascade* code behaves correctly in comparison to the analytical calculation of detection probability. To perform the study, a certain number of particles are thrown on the detector with a vertical zenith angle ($\theta = 0°$), for an energy of E= 114 MeV. The election of the energy is arbitrary but is of the order of particles expected to be detected by BATATA. For comparing the detection probability, the efficiency, i. e., the ratio of the number of particles reaching the detector and the total number of particles thrown, is considered as the probability.

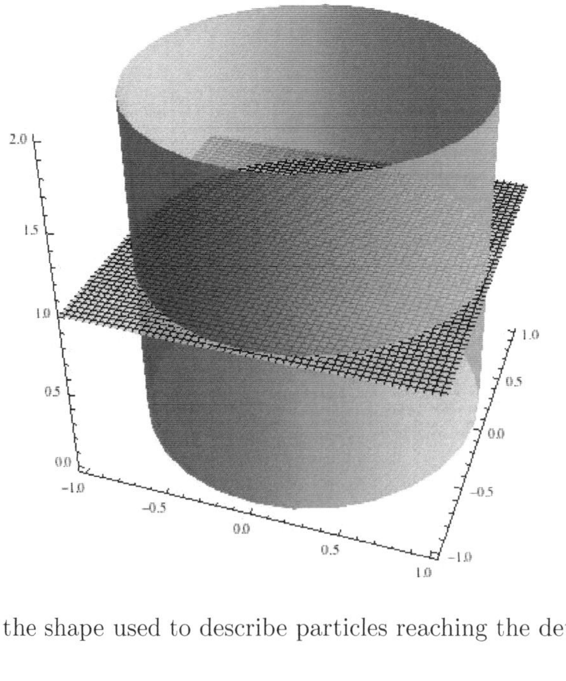

Figure 3. Schema of the shape used to describe particles reaching the detector, as used in the *cascade* code.

3.1. Equivalence of positions

For comparing the positions of the two codes (*cascade* and the analytical calculation) we should take into account the general characteristics of the detector. Figure 4 shows 11 positions along a diagonal of the detector, wich are considered as locations to throw the particles on, for a first study. These particles represent the detected particles.

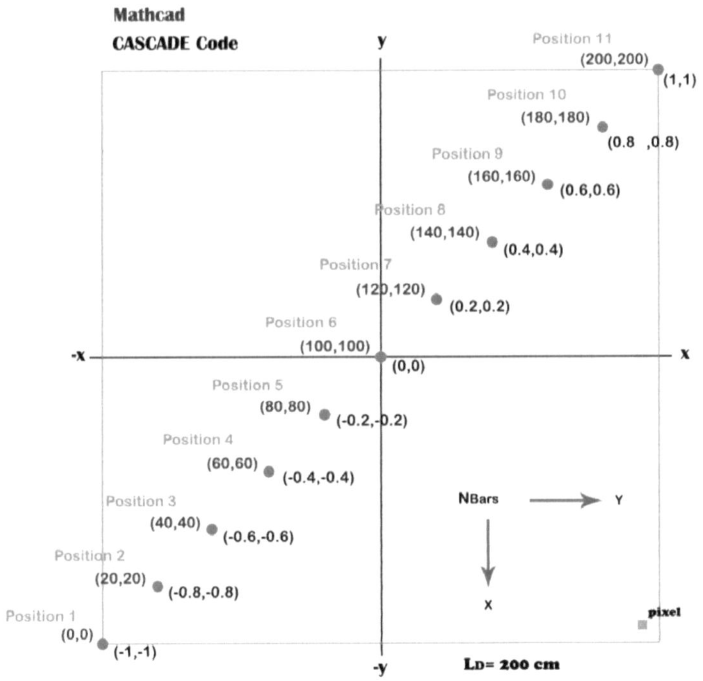

- N_{Bars}= 50
- x-Direction: i:1,2...N_{Bars}
- y-Direction: j:1,2...N_{Bars}
- Lateral side of the detector: L_D : 200 cm
- Pixel size: $\frac{L_D}{N_{bars}}$

Figure 4. Equivalence of the 11 positions in Mathcad and *cascade*. Blue points correspond to the location according to Mathcad coordinates and black points correspond to the *cascade's* ones.

3.2. Efficiency calculation

Cascade efficiency corresponds to the detection probability in BATATA, calculated in Mathcad.

Three cases are analyzed:

(a) 25,000 particles with $0° \leq \theta \leq 36°$, E= 114 MeV, $0° \leq \phi \leq 36°$ for each position of the diagonal were thrown (See Figure 5).

(b) From the 25,000 particles thrown in (a), only 1000 with theta between 6° and 21° are chosen. This is done for approaching as much as possible a vertical position, close to zero, according to the possibilities of the code. The efficiency is calculated for those particles.

IOP Conf. Series: Journal of Physics: Conf. Series **866** (2017) 012005 doi:10.1088/1742-6596/866/1/012005

(c) With the aim of obtaining a more accurate comparison, as is done for the analytical calculations (for **Mathcad** the angles considered are fixed values not ranges as in *cascade*), the range of angles is reduced to $3° \leq \theta \leq 9°$ and $33° \leq \phi \leq 39°$ (See Figure 5).

4. Results

4.1. Muon component

Figure 5 shows that the best option to check the compatibility between cascade and the analytical calculation is (c). The differences between the analytical calculations and simulations are in most of the locations, below 10%.

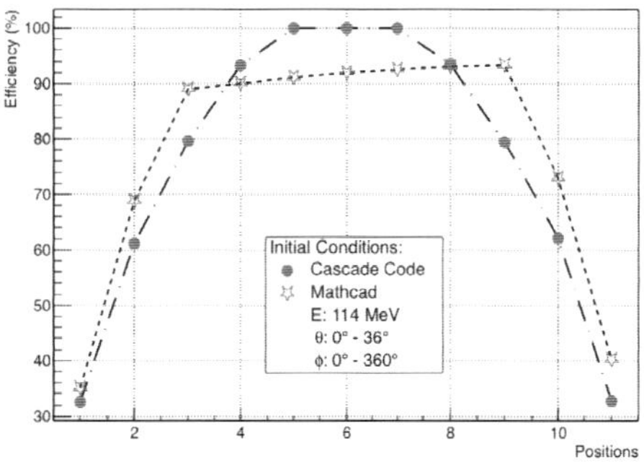

(a)

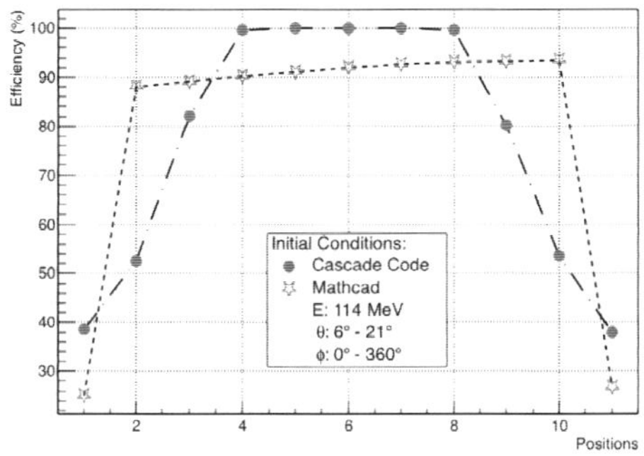

(b)

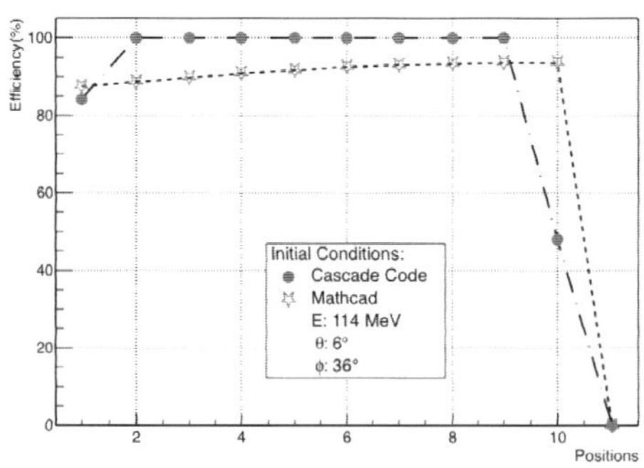

(c)

Figure 5. Comparison of compatibility between the code *cascade* and the analytical calculation described in Mathcad for the muon component.

4.2. Electromagnetic component

The results are shown in Figure 6. From these Figures it can be seen that *cascade* does not reproduce the analytical calculations for the electromagnetic component. It has been found out that *cascade* code does not includes the corresponding propagation equations for the electromagnetic component.

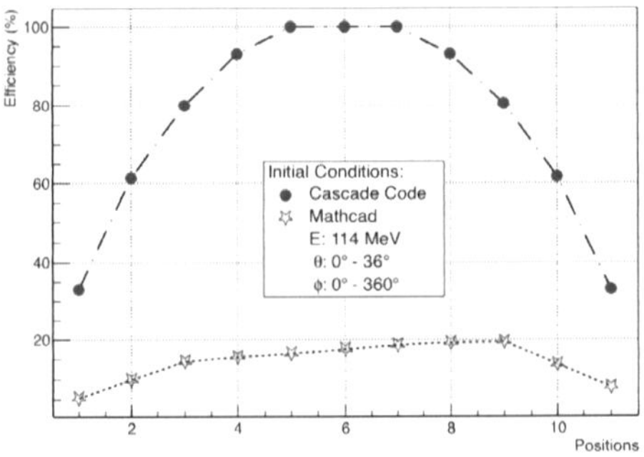

(a)

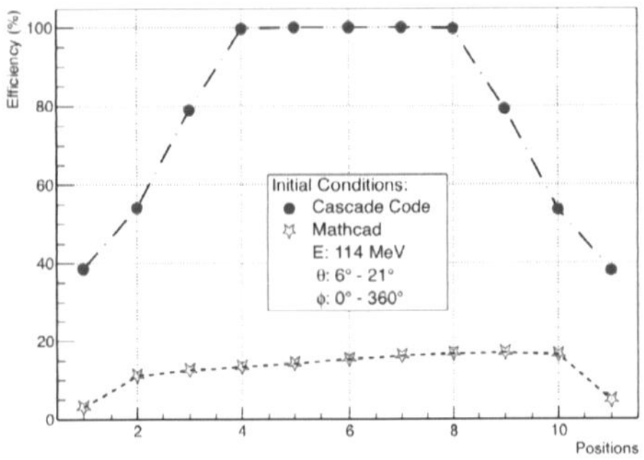

(b)

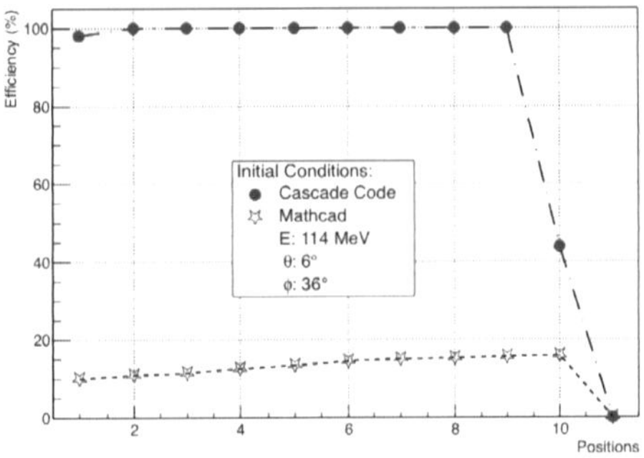

(c)

Figure 6. Comparison of compatibility between the code *cascade* and the analytical calculation described in Mathcad for the electromagnetic component.

5. Conclusions

The code *cascade* developed for describing measurements done by BATATA has been tested. The best initial conditions to use it for reproducing analytical calculations of efficiency for vertical zenith angles are found. At the moment *cascade* works properly for the muon component, however it can be adapted to include the code corresponding to the electromagnetic one.

6. Acknowledgments

1.- The Faculty of Science in Physics and Mathematics FCFM-UNACH.

2.- CONACyT for the support trough the Mesoamerican Center for Theoretically Physics and also for the project CB 243290.

3.- L'ORÉAL Mexico, CONACyT, UNESCO, CONALMEX and AMC for the support provided through the scholarship for Women in Science to Karen Salomé Caballero Mora.

4.- To Enrique Escalante for having provided the simulation code, Institute of Physics, UNAM.

References

[1] The Pierre Auger Collaboration, NIM A 798 (2015) 172-213, The Pierre Auger Cosmic Ray Observatory.

[2] The Pierre Auger Collaboration, arXiv:1604.03637v1, astroph-IM, The Pierre Auger Observatory Upgrade-Preliminary Design Report (2016).

[3] Medina Tanco G. et al. BATATA: A device to characterize the punch-through observed in underground muon detectors and to operate as a prototype for AMIGA, 2009, (arXiv:0909.3754v1).
[4] S.Riggi, A. Insolia, G. Medina-Tanco, E. Trovato, Reconstruction of muon tracks in a buried plastic scintillator muon telescope (BATATA), Nucl. Instr. Meth. A688 (2012) 22.
[5] The authors of this code are: Leonardo A. Castro Institute of Physics, Department of Theoretical Physics, UNAM; Jorge D. Brown of the Institute Nuclear Sciences, department of high energy physics, and Enrique Escalante, Institute of Physics, Department of Theoretical Physics, UNAM.
[6] I. A. Martnez Salazar. Caracterización de simulaciones de muones para el detector de centelleo BATATA, ubicado en el Observatorio Pierre Auger. Universidad Autnoma de Chiapas, Tesis de Licenciatura, 2016.

Photon emission and decay from generic Lorentz invariance violation

H. Martínez-Huerta and A. Pérez-Lorenzana

E-mail: hmartinez@fis.cinvestav.mx and aplorenz@fis.cinvestav.mx
Physics Department, Centro de Investigacion y de Estudios Avanzados del I.P.N., Apartado Postal 14-740, 07000, Mexico City, Mexico.

Abstract.
One of the most studied approaches in phenomenology to introduce the breaking of Lorentz symmetry is the generic approach. This consist on the modification of the free particle dispersion relation by the addition of an extra power law term of order n on energy or momentum. Using this approach in the photon sector, we have calculated the generic rates for vacuum Cherenkov radiation and photon decay, for any order n, at leading order. Explicit results for the decay and emission rates for the lowest values of n are also presented.

1. Introduction

Lorentz invariance violation (LIV), mainly motivated by quantum gravity and string theories [1, 2, 3, 4], represents an interesting sector in the search for physics beyond the Standard Model. Since Lorentz symmetry has a fundamental role in the construction of the model, the derived physics form LIV extensions tends to be unique and energy dependent in order to preserves the standard physics unchanged [5, 6]. Additionally, LIV evidence is usually expected to increase with the energy and distance.

Currently, the most energetic known phenomena are cosmic rays (CR). They can reach energies of several decades of EeV [7, 8] and can travel astrophysical distances before detection. Moreover, their study has shown a strong development in recent decades. Therefore, properly studied LIV consequences could be identified in the current CR observatories and experiments.

Accordingly to the previous ideas, in this note we present a phenomenological first order correction, using a generic approach, of the emission rates for two processes that could have a significant impact on cosmic particle propagation: photon decay and vacuum Cherenkov radiation.

The generic mechanism for the introduction of LIV, frequently used in the literature [5, 6, 9, 10, 11, 12, 13, 14], converges to an explicit not Lorentz invariant term added to the free particle Lagrangian density that will generate a correction to the dispersion relation

$$S_a = E_a^2 - p_a^2 = m_a^2 \pm \alpha_{a,n} A^{n+2}, \tag{1}$$

where E_a and p_a stand for the four momenta associated with an a particle species. A can take the form of E or p for specific models, nevertheless, for $m_a \ll \{E, p\}$, the ultrarelativistic limit, any particular choice of A will be equivalent. The coefficient $\alpha_{a,n}$ in Eq. (1), parametrizes the particle dependent LIV correction, where n expresses the correction order to the mass shell codified in the energy or momentum expression in A. The generic α_n is frequently inversely related to n-th power of E_{QG}, the scale of quantum gravity (QG) or the scale of the new physics beneath, which is expected to be close to 10^{19} GeV. Several methods are used in the search for LIV signals, some of them can lead to lower limits to E_{QG} [15, 16, 17, 18, 19, 20, 21].

In the next section, we have applied the LIV generic correction in Eq. (1) for photons to find photon emission and decay rate for the processes depicted in Figure 1 and 2, corrected at first α_n order but for any n. Such processes are forbidden in the standard theory by energy momentum conservation, but under the LIV hypothesis they can be permitted and can be used in the search for LIV signatures. In Figure 1, we show the emission of a single photon by a charged particle that propagates in vacuum, commonly named vacuum Cherenkov radiation [5, 11, 12] due its role as an energy loss mechanism in charged particle propagation. Figure 2 shows LIV photon decay [5, 11, 13], this process is kinematically allowed and motivated by the LIV extra term in Eq. (1). The simplest process will produce an electron - positron pair. Once the general expression for the corrected rates had been presented, we shall used them to show their behaviour for particular values of n.

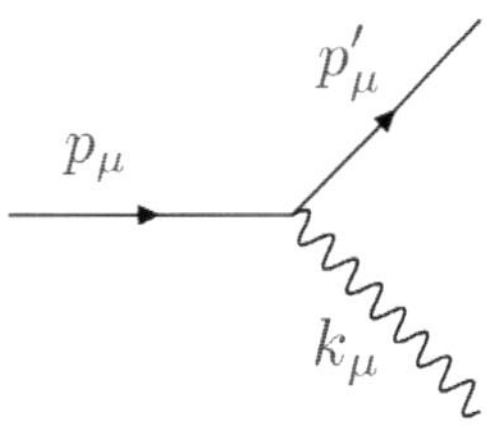

Figure 1. Diagram for LIV vacuum Cherenkov radiation.

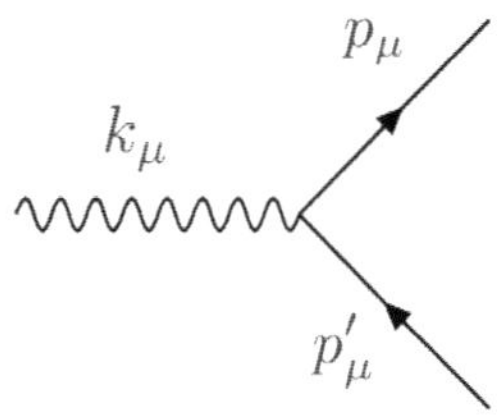

Figure 2. Diagram for LIV photon decay.

2. LIV corrected rates

Following the diagrams in Figures 1 and 2, photon four momenta will be named $(\omega, \mathbf{k})$, whereas prime and not prime notation stand for the fermion momenta, as indicated. A proton would be assumed for the first and an $e^+ - e^-$ pair for the last. They have been chosen due their relevance for cosmic [7, 22] and gamma rays [23]. Hence, the corrected square amplitude at first LIV order in α_n will be given by:

$$\frac{1}{2}\sum_{spin}|M|^2 = e^2|4m_a^2 - \alpha_n k^{n+2}|, \tag{2}$$

where orthogonality relation is assumed for photon polarizations at first approximation, and $k = |\mathbf{k}|$. The absolute value above is taking in order to ensure physical congruence due the limits and the generality in the sign of α_n. For a more detailed calculation see Ref. [24].

Using Eq. (2) and the LIV corrected energy for photons, $\omega(k, \alpha_n) = k\sqrt{1+\alpha_n k^n}$, we derived the modified decay and emission rates on an arbitrary preferential frame for any order n. For vacuum Cherenkov radiation, we found that

$$\Gamma^{(n)}_{a\to a\gamma} = \frac{e^2}{4\pi}\frac{1}{4E_a}\int_0^{\theta}\sum_{k_i}\frac{|4m_a^2 - \alpha_n k_i^{(n+2)}|}{\omega(k_i,\alpha_n)}\frac{k_i^2\sin\theta d\theta}{|p_a\cos\theta - k_i - \frac{(1+\frac{2+n}{2}\alpha_n k_i^n)}{\sqrt{1+\alpha_n k_i^n}}\sqrt{k_i^2 + E_a^2 - 2k_i p_a\cos\theta}|}, \tag{3}$$

where k_i are the non zero photon momenta modes from the corrected energy-momenta conservation. For $n = 0$ we have:

$$k_0^{(n=0)} = \frac{1}{p\cos\theta}\left[E^2(\alpha_0 + 1) - p^2\cos^2\theta\right], \tag{4}$$

and for $n = 1$,

$$k_\pm^{(n=1)} = \frac{1}{2p_a\cos\theta}\left(E_a^2 \pm \sqrt{E_a^4 + \frac{4p_a\cos\theta}{\alpha_1}(E_a^2 - p_a^2\cos^2\theta)}\right). \tag{5}$$

Since k_- is non physical, it will not be used in our phenomenological approach.

Likewise, for LIV photon decay of order n we get the rate

$$\Gamma^{(n)}_{\gamma\to e^+e^-} = \frac{e^2}{4\pi}\frac{|4m_e^2 - \alpha_n k^{n+2}|}{4\omega(k,\alpha_n)}\int_0^\pi \sum_{p=p_\pm} \frac{p^2\sin\theta d\theta}{|(k\cos\theta - p)E_e - p\sqrt{k^2 + E_e^2 - 2kp\cos\theta}|}, \tag{6}$$

where the momenta modes from the corrected energy-momenta conservation and for any n are

$$p_\pm = \frac{1}{2(\alpha_n k^n + \sin^2\theta)}\left(\alpha_n k^{n+1}\cos\theta \pm \sqrt{\alpha_n^2 k^{2n+2}\cos^2\theta - 4(\sin^2\theta + \alpha_n k^n)(1 + \alpha_n k^n)m^2}\right). \tag{7}$$

Here, e stands for the electron charge, θ for the angle between final particles and $p_a = \sqrt{E_a^2 - m_a^2}$ for the given particle a absolute momentum. Photon decay expression are generic for any fermion pair in the final state, provided the corresponding mass is used, while the emission rate is, from an phenomenological approach, generic for any charged particle or nucleus with spin $= 1/2$. There exist vacuum Cherenkov radiation and photon decay rates obtained from different LIV approaches. For instance, expressions from the minimal Standard-Model extension with spontaneous breaking of Lorentz symmetry [25] and from the introduction of Lorentz violating operators of dimensions four and six can be found in Refs. [26, 27].

Integration on Eqs. (3) and (6), for different n, were numerically performed. The results are depicted in Figures 3 and 4 for Vacuum Cherenkov radiation and LIV photon decay respectively. A small angle approach is taken for vacuum Cherenkov radiation in order to clarify the threshold. Results for fixed n and different α_1 for both processes can be found in [24].

As it can be seen in Figs. 3 and 4, the phenomena are sensitive to the LIV term. Both rates grow with the primary particle energy and they show a different slope with the order of the correction n. For vacuum Cherenkov emission a phenomenological threshold is required to turn off the process and preserve the physics below it unaffected. We have proposed $E \geq (m^2/\alpha_0)^{1/2}$ for $n = 0$ and $E \geq (4m^2/\alpha_1)^{1/3}$ for $n = 1$. In LIV photon decay case, there is also a threshold that turns off the process at low energies but it naturally comes from energy-momentum conservation in the Lorentz conserving fermion sector and the corrected LIV photon.

3. Conclusions

We have found a generic first order LIV correction for every order n to the emission and decay rates for vacuum Cherenkov radiation and LIV photon decay. Unlike Lorentz invariant theory, both processes are permitted under LIV hypothesis. As expected, their possible consequences increases with the energy. They show the same phenomenological behaviour at the first orders in n and can lead to different expected physics at the most energetic scenarios, such as cosmic rays, but we will present such an analysis in a future work.

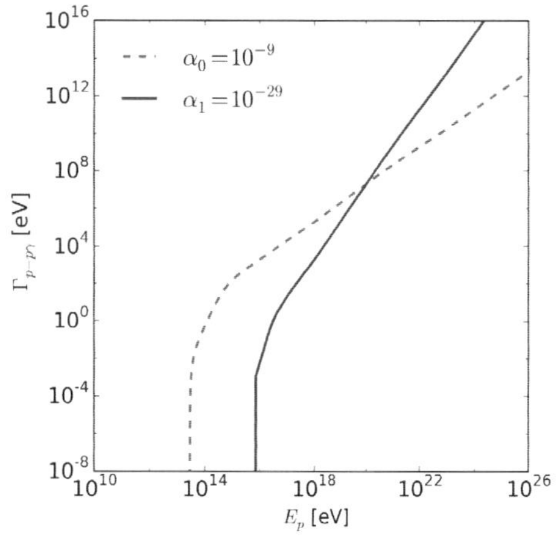

Figure 3. Emission rate for LIV vacuum Cherenkov radiation for n=0,1, $m_a = m_{proton}$ and $\theta_{max} = 10^{-2}$.

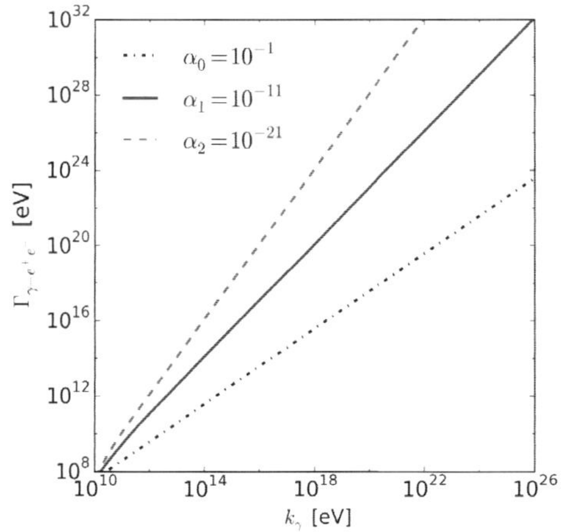

Figure 4. Decay rate for LIV photon decay into electron positron pairs, for n=0,1,2.

Acknowledgments
This work was partially supported by Conacyt grant No. 237004. We are also grateful to Conacyt for the support trough the Mesoamerican Center for Theoretically Physics.

References

[1] Ellis J, Mavromatos N E and Nanopoulos D V 2000 *Phys. Rev.* D **61** 027503
[2] Gambini R and Pullin J 1999 *Phys. Rev.* D **59** 124021
[3] Ellis J, Mavromatos N and Nanopoulos D 2000 *Gen. Rel. Grav.* **32** 127–144
[4] Ellis J, Mavromatos N, Nanopoulos D and Volkov G 2000 *Gen. Rel. Grav.* **32** 1777 – 1798
[5] Coleman S and Glashow S L 1999 *Phys. Rev.* D **59** 116008
[6] Amelino-Camelia G, Ellis J, Mavromatos N E, Nanopoulos D V and Sarkar S 1998 *Nature* **393** 763–765
[7] Aab A *et al.* (Pierre Auger Collaboration) 2014 *Phys. Rev.* D **90** 122005
[8] Abraham J *et al.* (Pierre Auger Collaboration) 2010 *Phys. Lett.* B **685** 239–246
[9] Amelino-Camelia G 2001 *Nature* **410** 1065–1067
[10] Ahluwalia D V 1999 *Nature* **398** 199
[11] Galaverni M and Sigl G 2008 *Phys. Rev. Lett.* **100** 021102
[12] Gagnon O and Moore G D 2004 *Phys. Rev.* D **70** 065002
[13] Galaverni M and Sigl G 2008 *Phys. Rev.* D **78** 063003
[14] Martínez H and Pérez-Lorenzana A 2013 *Journal of Physics: Conference Series* **468** 012005
[15] Ellis J R, Mavromatos N E, Nanopoulos D V and Sakharov A S 2003 *Astron. Astrophys.* **402** 409–424
[16] Albert J *et al.* (MAGIC, Other Contributors) 2008 *Phys. Lett.* **B668** 253–257 (*Preprint* 0708.2889)
[17] Ellis J R, Mavromatos N E, Nanopoulos D V, Sakharov A S and Sarkisyan E K G 2006 *Astropart. Phys.* **25** 402–411 [Erratum: Astropart. Phys.29,158(2008)] (*Preprint* 0712.2781)
[18] Abramowski A *et al.* (HESS Collaboration) 2011 *Astropart. Phys.* **34** 738747
[19] Vasileiou V, Jacholkowska A, Piron F, Bolmont J, Couturier C, Granot J, Stecker F W, Cohen-Tanugi J and Longo F 2013 *Phys. Rev.* D **87** 122001
[20] Xu H and Ma B Q 2016 *Phys. Lett.* B **760** 602–604
[21] Nellen L (HAWC) 2015 *Proc. Sci., ICRC2015 (2016) 850* (*Preprint* 1508.03930)
[22] Aab A *et al.* (Pierre Auger Collaboration) 2014 *Phys. Rev.* D **90** 122006
[23] De Angelis A, Galanti G and Roncadelli M 2013 *Mon. Not. R. Astron. Soc.* **432** 3245–3249
[24] Martńez-Huerta H and Pérez-Lorenzana A 2016 *Journal of Physics: Conference Series* **761** 012035
[25] Colladay D and Kostelecky V A 1998 *Phys. Rev. D* **58** 116002
[26] Klinkhamer F R and Schreck M 2008 *Phys. Rev. D* **78** 085026
[27] Grigory Rubtsov P S and Sibiryakov S 2012 *Phys. Rev. D* **86** 085012

IOP Conf. Series: Journal of Physics: Conf. Series **866** (2017) 012007 doi:10.1088/1742-6596/866/1/012007

Dark Matter Indirect Detection with Gamma Rays

J. Patrick Harding

Physics Division, Los Alamos National Laboratory, Los Alamos, NM, USA

E-mail: jpharding@lanl.gov

Abstract. Searches for weakly interacting massive particle (WIMP) dark matter with gamma-ray instruments are a way to get a unique observational handle on the particle nature of dark matter. I will discuss the details of how to perform these searches, both for annihilating and decaying WIMPs. I will discuss the calculation of the gamma-ray flux from possible sources of dark matter annihilation or decay and show examples of limits which have been calculated using these techniques.

1. Evidence for Dark Matter

Since its discovery in the 1930s, the evidence for dark matter has been plentiful. Dark matter was first discovered by observations that rotation curves in galaxies and clusters differed significantly from the rotation curve expected if the gravitational potential was dominated by the observed luminous matter [1]. The luminous matter of galaxies is largely contained within the inner few kiloparsecs of the galaxy. According to the law of gravitation, in the outer regions of a galaxy, the circular velocity of observed bodies should therefore follow:

$$F_G = \frac{GmM(<r)}{r^2} = \frac{mv_{\rm cir}^2}{r} \quad (1)$$

$$v_{\rm cir}(r) \propto \sqrt{\frac{M(<r)}{r}} , \quad (2)$$

with $M(<r)$ the mass contained with radius from the galactic center r and $v_{\rm cir}$ the circular velocity at radius r. G is the universal gravitational constant and m is a test mass, which cancels out in the second equation. However, the circular velocity tends to flatten out at large radii rather than decreasing, consistent with

$$\begin{aligned} M(<r) &\propto r \\ \rho_M &\propto r^{-2} . \end{aligned} \quad (3)$$

This indicates an additional mass component to galaxies which is not luminous and has a density profile $\rho_M \sim r^{-2}$ which continues out to large radii. This additional, non-luminous mass is referred to as "dark matter". Further evidence for the dark matter comes from anisotropies in the cosmic microwave background (CMB), which has shown that the dark matter makes up 26.7% of the energy budget of the universe, but ordinary baryonic matter makes up only 4.9% [2]. The case for dark matter has also been made based on large-scale structure of the universe [3, 4] and observations of galaxy clusters [5, 6].

Although dark matter has only been observed through its gravitational interactions, the properties of dark matter are well constrained. Lack of sufficient gravitational microlensing have determined that the dark matter is made up of particles with masses $M_\chi \lesssim 10^{-7} M_\odot$, and baryonic clumps of matter cannot explain the majority of the dark matter observed in the Galaxy [7, 8, 9].

The most popular dark matter model is the weakly interacting massive particle (WIMP), with a dark matter particle of mass $M_\chi \sim 100$ GeV which interacts through a force of similar strength to the weak force. WIMPs were formed in the early universe in equilibrium with other particles at high temperature. As the universe cooled, WIMPs collided and annihilated into standard model particles. However, once the expansion of the universe exceeded the annihilation rate of the WIMPs, they stopped annihilating and the number density of WIMPs "froze out", giving the dark matter number density seen today. The so-called "WIMP Miracle" is that for a WIMP with weak-scale cross-section ($\sigma \sim 1$ pb) and a weak-scale mass ($M_\chi \sim 100$ GeV), the relic density of thermal WIMPs is $\Omega_{WIMP} \approx 0.2$, consistent with the energy density for the dark matter. A detailed calculation of the thermal cross-section is found in Ref. [10].

Though the WIMP model is often motivated by Supersymmetry, other theories of beyond-the-Standard-Model physics predict WIMPs as well, such as models with compact extra dimensions. Most of the current models predict WIMP masses from 10 GeV-5 TeV, but WIMPs as heavy as several PeV could exist as well. The typical velocity-weighted WIMP annihilation cross-section is $\langle\sigma v\rangle = 2.2^{-26}\, cm^3/s$ [10]. If the WIMPs are not produced as thermal relics but are produced non-thermally though decays of heavier particles, or if the cross-section is boosted through Sommerfeld enhancement (discussed in Section 3), then the cross-section today could be much different from this value [11].

2. Calculations of the Dark Matter Annihilation Flux

2.1. Particle Flux from Dark Matter Annihilations

The number of dark matter annihilations per unit volume is $\rho(\vec{x})^2/(2M_\chi^2)\langle\sigma_{\rm A} v\rangle$, for dark matter mass M_χ and dark matter number density ρ/M_χ . $\langle\sigma_{\rm A} v\rangle$ is the annihilation cross-section times the relative velocity between interacting dark matter particles, averaged over their velocity distribution, and the factor of two comes from symmetry concerns. The number of dark matter collisions per unit time per unit observational area integrated along a particular line-of-sight is then

$$\frac{d^2N}{dtdA} = \frac{\langle\sigma_{\rm A} v\rangle}{2M_\chi^2} \iint \frac{r^2 dr d\Omega}{4\pi r^2} \rho(\vec{x})^2 = \frac{\langle\sigma_{\rm A} v\rangle}{2M_\chi^2} \iint \frac{dr d\Omega}{4\pi} \rho(\vec{x})^2 \ . \tag{4}$$

Observationally, we detect the number of Standard Model particles per unit energy produced by dark matter annihilations rather than the annihilations themselves. For an annihilation final-state f, the branching fraction Br_f is the fraction of time a dark matter annihilation will end up in that final state. For a particular Standard Model particle i, $dN_f^{(i)}/dE$ is the number of particles i per unit energy produced in final-state f. The details of which final states occur with which branching fractions depends on the precise dark matter model under consideration, though often, for simplicity, a particular dark matter annihilation channel is assumed to dominate and its branching fraction is set to unity. To simulate hadronization and decay of standard model particles into final states, analytical expressions or simulations are used (see section 3.1). The flux of gamma rays per units energy and solid angle coming from a region of dark matter annihilations is given by

$$\frac{d^2F_\gamma}{dEd\Omega} = \frac{\langle\sigma_{\rm A} v\rangle}{2} \sum_f \frac{dN_f^{(\gamma)}}{dE} \mathrm{Br}_f \int_0^\infty \frac{dr}{4\pi} \frac{\rho(\vec{x})^2}{M_\chi^2} \ . \tag{5}$$

Here, r is the integration over the line-of-sight toward the source. $|\vec{x}| = \sqrt{R^2 - 2rR\cos(\theta) + r^2}$ is the distance from a differential volume of the source to the source center, where R is the distance from the Earth to the source and θ is the observation angle from the source center to the differential volume.

2.2. *Dark Matter Distributions*

Dark matter in galaxies exists in a roughly spherically symmetric "halo" that approximately falls off as $\rho(|\vec{x}|) \propto |\vec{x}|^{-2}$ with the distance from the galactic center. Numerical N-body simulations have been done considering both pure dark matter and dark matter with baryons [12, 13, 14, 15, 16, 17, 18, 19, 20, 21, 22, 23, 24]. The most widely adopted dark matter profile is the Navarro-Frenk-White (NFW) profile [18], parameterized as

$$\rho(|\vec{x}|) = \frac{\rho_s}{(|\vec{x}|/r_s)(1 + |\vec{x}|/r_s)^2} \tag{6}$$

with scale radius $r_s \approx 20$ kpc for our Galaxy. A somewhat shallower dark matter profile, the Einasto profile, is parameterized as [25, 26]

$$\rho(|\vec{x}|) = \rho_s \exp\left[-\frac{2}{\alpha}\left(\left(\frac{|\vec{x}|}{r_s}\right)^{\alpha} - 1\right)\right], \tag{7}$$

with scale radius $r_s \approx 20$ kpc and $\alpha = 0.1 - 0.2$ for Milky-Way-type halos [19, 23]. The scale density for these Galactic dark matter profiles, ρ_s, is fixed such that the total integrated dark matter density of the object gives its measured dark matter mass.

3. Cross-section and Substructure Boosts

The dark matter cross-section can be larger than thermal if the dark matter couples to light gauge bosons. The exchange of the bosons by the annihilating dark matter particles can create a resonance and increase the annihilation cross-section today by orders of magnitude relative to the value at the time of freeze-out. This is called "Sommerfeld enhancement", after a similar effect due to photons in electron scattering. Sommerfeld enhancement goes as [27]

$$\epsilon_v \equiv \frac{v}{\alpha} \tag{8}$$

$$\epsilon_\phi \equiv \frac{m_\phi}{\alpha M_\chi} \tag{9}$$

$$S \approx \frac{\pi}{\epsilon_v} \frac{\sinh\left(\frac{2\pi\epsilon_v}{\pi^2\epsilon_\phi/6}\right)}{\cosh\left(\frac{2\pi\epsilon_v}{\pi^2\epsilon_\phi/6}\right) - \cos\left(2\pi\sqrt{\frac{1}{\pi^2\epsilon_\phi/6} - \frac{\epsilon_v^2}{(\pi^2\epsilon_\phi/6)^2}}\right)}, \tag{10}$$

where $\alpha \sim 1/35$ is the weak coupling, m_ϕ is the gauge boson mass, M_χ is the dark matter mass, v is the relative velocity of the dark matter particles, and S is the enhancement to the cross-section. Because the Sommerfeld enhancement increases at low velocities, and the relative velocity of dark matter in the Galaxy is $\sim 300\,km\,s^{-1}$, much smaller than the $\sim c/4$ velocities in the early universe, the dark matter cross-section during thermal freeze out can be much less than the cross-section today. The shape of Sommerfeld enhanced cross-section limits have multiple peaks in them, corresponding to the combinations of velocity + dark matter mass + gauge boson mass which lie directly at resonance, not merely near it. Usually, Sommerfeld enhancement is achieved using a dark-sector light boson. However, for the heavy dark matter masses relevant to HAWC, Standard Model W and Z bosons are light enough to cause this enhancement. Especially for WIMPs which annihilate primarily into W^+W^- or Z^+Z^- states, they are guaranteed to couple to Standard Model gauge bosons and naturally have Sommerfeld enhancement.

The flux can also be boosted due to dark matter substructure. In all numerical simulations of dark matter, most of the dark matter is not in a smooth halo but is found in smaller subhalos, which can have masses as low as $10^{-6}M_{\odot}$. Because the flux is proportional to the dark matter density squared, and the local density in these subhalos is much larger than from the smooth halo, these substructures can increase the dark matter flux from 2-1000 times, depending on the object. Most of the subhalos are far from the center of the smooth halo [28]. Therefore, in observations of the Galactic center, substructure boosts do not contribute to the flux significantly. For dwarf galaxies, which are actually large, virialized substructures themselves, the substructure boost does not contribute much because the objects are small in extent and do not have that many orders of magnitude difference between the smooth halo mass and the size of the substructure. However, for extragalactic dark matter sources, especially other galaxies and galaxy clusters, the substructure boost can be large.

3.1. Calculation of Dark Matter Spectra

To calculate the photon spectrum for a particular WIMP annihilation channel, programs like PYTHIA are used to simulate the photon radiation of charged particles as well as decays of particles such as the π^0 [29]. Because even fairly stable particles decay within a short distance on astrophysical scales, all possible particles are decayed, including pions, muons, and neutrons. Using a large sample of events for each final state and each value of M_χ, the number of photons in the final state in a given logarithmic energy bin is counted and averaged over the number of events, yielding the average number of photons in that energy bin per annihilation event.

4. Dark Matter Decay

The procedures for getting dark matter limits from decay are very similar to those from dark matter annihilation, with a few small changes. First, the center-of-mass energy available for decay channels in the dark matter mass, M_χ for decay, instead of $2M_\chi$ in annihilation. This is because a decaying particle has only one particle-mass to covert into final states instead of two. Second, the source shape and extension for sources of dark matter decay are different than for annihilation. The flux for annihilation goes as the density squared (ρ^2) whereas for decay the flux goes as the density (ρ). Third, for dark matter annihilation, the limit is an *upper* limit on the velocity-weighted dark matter cross-section $\langle\sigma v\rangle$. For decay, it is a *lower* limit on the dark matter lifetime τ. These differences are discussed in more detail below.

4.1. Particle Flux from Dark Matter Annihilations

The number of dark matter decays per unit volume per unit time is $\rho(\vec{x})/(M_\chi\tau)$, for dark matter mass M_χ and dark matter number density ρ/M_χ . τ is the dark matter lifetime, which must be much longer than the age of the universe. The number of dark matter collisions per unit time per unit observational area integrated along a particular line-of-sight is then

$$\frac{d^2N}{dtdA} = \frac{1}{M_\chi\tau}\int\frac{r^2drd\Omega}{4\pi r^2}\rho(\vec{x}) = \frac{1}{M_\chi\tau}\int\frac{drd\Omega}{4\pi}\rho(\vec{x}) \ . \quad (11)$$

Note that for decay, the total center-of-mass energy is M_χ, rather than $2M_\chi$ for annihilation. This yields a flux equation similar to that for annihilation, but depending linearly on the dark matter density ρ instead of quadratically. The flux of gamma rays per units energy and solid angle coming from a region of dark matter annihilations is given by

$$\frac{d^2F_\gamma}{dEd\Omega} = \frac{1}{\tau}\sum_f\frac{dN_f^{(\gamma)}}{dE}\mathrm{Br}_f\int_0^\infty\frac{dr}{4\pi}\frac{\rho(\vec{x})}{M_\chi} \ . \quad (12)$$

IOP Conf. Series: Journal of Physics: Conf. Series **866** (2017) 012007 doi:10.1088/1742-6596/866/1/012007

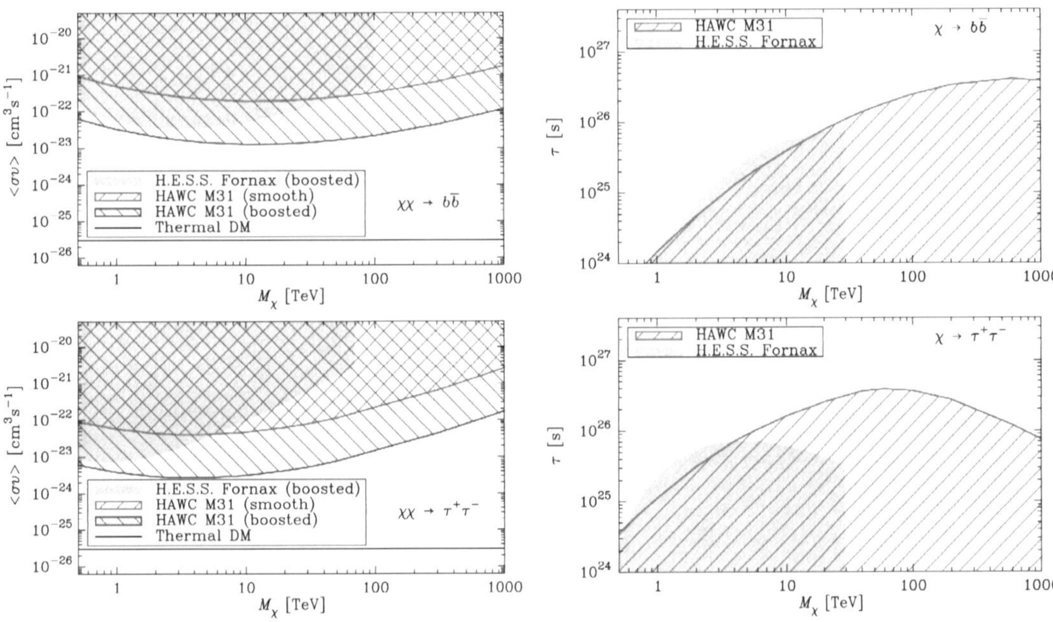

Figure 1. Sample dark matter annihilation and decay limit plots for two dark matter channels, bottom quarks and tau leptons. Information about the data on these figures can be found in Ref. [30].

The dark matter spatial distributions for decaying dark matter are the same as for annihilating dark matter. Because dark matter decay involves only one dark matter particle, enhancements such as Sommerfeld enhancement do not apply. Additionally, because the flux goes as the density rather than the density squared for decay, dark matter substructure does not give additional boosts to the flux over the smooth dark matter halo.

5. Calculating Dark Matter Limits

To get limits on the WIMP annihilation cross-section (decay lifetime), the dark matter annihilation (decay) channel and the dark matter mass are set to fixed values. This leaves the annihilation cross-section (decay lifetime) as the only free parameter in the flux. A χ^2 value or likelihood test-statistic is calculated which compares a model with the dark matter gamma-ray source to a model without it. The flux is scaled by changing the cross-section or lifetime until the χ^2 (test statistic) is minimized (maximized). Then the value of the annihilation cross-section (decay lifetime) is increased (decreased) until the χ^2 or test statistic changes from its optimal value by 2.71 [31]. This is the value of the cross-section (lifetime) which is constrained at 95% confidence level, which is the standard for dark matter limits. This process is repeated for all dark matter masses and channels that are being considered. The results are then plotted as several log-log plots, one for each dark matter channel, with the dark matter mass on the x-axis and the cross-section (lifetime) on the y-axis. Some example figures from Ref. [30] are shown as Figure 1 for illustration. Note that all cross-section *above* the exclusion line are excluded while all decay lifetimes *below* the exclusion line are excluded. It is also common to show limits from more than one experiment on the same figure for comparison or to show expected values from dark matter models (e.g. the thermal dark matter cross-section) on these figures as well. In this way, you can tell the relevance of a particular experiment to dark matter searches.

Acknowledgments
We acknowledge support from the US Department of Energy Office of High-Energy Physics and The Laboratory Directed Research and Development (LDRD) program of Los Alamos National Laboratory. We are grateful to Conacyt for the support through the Mesoamerican Center for Theoretical Physics.

References

[1] Zwicky F 1933 *Helv.Phys.Acta* **6** 110–127
[2] Ade P A R *et al.* (Planck) 2014 *Astron. Astrophys.* **571** A16 (*Preprint* 1303.5076)
[3] Larson D, Dunkley J, Hinshaw G, Komatsu E, Nolta M *et al.* 2011 *Astrophys.J.Suppl.* **192** 16 (*Preprint* 1001.4635)
[4] Reid B A, Percival W J, Eisenstein D J, Verde L, Spergel D N *et al.* 2010 *Mon.Not.Roy.Astron.Soc.* **404** 60–85 (*Preprint* 0907.1659)
[5] Gnedin O Y and Ostriker J P 2000 *Astrophys.J.* (*Preprint* astro-ph/0010436)
[6] Randall S W, Markevitch M, Clowe D, Gonzalez A H and Bradac M 2008 *Astrophys.J.* **679** 1173–1180 (*Preprint* 0704.0261)
[7] Alcock C *et al.* (MACHO) 1996 *Astrophys.J.* **471** 774 (*Preprint* astro-ph/9604176)
[8] Alcock C *et al.* (MACHO) 2000 *Astrophys.J.* **542** 281–307 (*Preprint* astro-ph/0001272)
[9] Wyrzykowski Ł, Kozłowski S, Skowron J, Udalski A, Szymański M K, Kubiak M, Pietrzyński G, Soszyński I, Szewczyk O, Ulaczyk K and Poleski R 2011 *Mon. Not. Roy. Astron. Soc.* **413** 493–508 (*Preprint* 1012.1154)
[10] Steigman G, Dasgupta B and Beacom J F 2012 *Phys.Rev.* **D86** 023506 (*Preprint* 1204.3622)
[11] Profumo S 2005 *Phys.Rev.* **D72** 103521 (*Preprint* astro-ph/0508628)
[12] Blumenthal G R, Faber S, Flores R and Primack J R 1986 *Astrophys.J.* **301** 27
[13] Diemand J *et al.* 2008 *Nature* **454** 735–738 (*Preprint* 0805.1244)
[14] Gnedin O Y, Kravtsov A V, Klypin A A and Nagai D 2004 *Astrophys. J.* **616** 16–26 (*Preprint* astro-ph/0406247)
[15] Gnedin O Y, Ceverino D, Gnedin N Y, Klypin A A, Kravtsov A V, Levine R, Nagai D and Yepes G 2011 *ArXiv e-prints* (*Preprint* 1108.5736)
[16] Guedes J, Callegari S, Madau P and Mayer L 2011 *Astrophys. J.* **742** 76 (*Preprint* 1103.6030)
[17] Macciò A V, Stinson G, Brook C B, Wadsley J, Couchman H M P, Shen S, Gibson B K and Quinn T 2012 *Astrophys. J. Lett.* **744** L9 (*Preprint* 1111.5620)
[18] Navarro J F, Frenk C S and White S D 1997 *Astrophys.J.* **490** 493–508 (*Preprint* astro-ph/9611107)
[19] Navarro J F, Ludlow A, Springel V, Wang J, Vogelsberger M *et al.* 2008 (*Preprint* 0810.1522)
[20] Pedrosa S, Tissera P B and Scannapieco C 2010 *Mon. Not. Roy. Astron. Soc.* **402** 776–788 (*Preprint* 0910.4380)
[21] Romano-Diaz E, Shlosman I, Hoffman Y and Heller C 2008 *Astrophys. J.* **685** L105 (*Preprint* 0808.0195)
[22] Romano-Diaz E, Shlosman I, Heller C and Hoffman Y 2009 *Astrophys. J.* **702** 1250–1267 (*Preprint* 0901.1317)
[23] Stadel J, Potter D, Moore B, Diemand J, Madau P, Zemp M, Kuhlen M and Quilis V 2009 *Mon. Not. Roy. Astron. Soc.* **398** L21–L25 (*Preprint* 0808.2981)
[24] Tissera P B, White S D M, Pedrosa S and Scannapieco C 2010 *Mon. Not. Roy. Astron. Soc.* **406** 922–935 (*Preprint* 0911.2316)
[25] Einasto J 1965 *Trudy Inst. Astrofiz. Alma-Ata* **51** 87
[26] Springel V, Wang J, Vogelsberger M, Ludlow A, Jenkins A *et al.* 2008 *Mon.Not.Roy.Astron.Soc.* **391** 1685–1711 (*Preprint* 0809.0898)
[27] Feng J L, Kaplinghat M and Yu H B 2010 *Phys.Rev.* **D82** 083525 (*Preprint* 1005.4678)
[28] Kamionkowski M, Koushiappas S M and Kuhlen M 2010 *Phys.Rev.* **D81** 043532 (*Preprint* 1001.3144)
[29] Sjostrand T, Mrenna S and Skands P Z 2006 *JHEP* **0605** 026 (*Preprint* hep-ph/0603175)
[30] Harding J P (HAWC) 2015 *Proc. 34th ICRC* (The Hague, The Netherlands)
[31] Ackermann M *et al.* 2014 *Physical Review D* **89** ISSN 1550-7998, 1550-2368

IOP Conf. Series: Journal of Physics: Conf. Series **866** (2017) 012008 doi:10.1088/1742-6596/866/1/012008

Astroparticle Physics Tests of Lorentz Invariance Violation

R G Lang and V de Souza

Instituto de Física de São Carlos, Universidade de São Paulo. Avenida Trabalhador São-Carlense, 400, São Carlos, SP, Brazil

E-mail: rodrigo.lang@usp.br

Abstract. Testing Lorentz invariance is essential as it is one of the pillars of modern physics. Moreover, its violation is foreseen in several popular Quantum Gravity models. Several authors study the effects of Lorentz invariance violation (LIV) in the propagation of ultra-high energy cosmic rays. These particles are the most energetic events ever detected and therefore represent a promising framework to test LIV. In this work we present an analytic calculation of the inelasticity for any $a + b \to c + d$ interaction using first order perturbation in the dispersion relation that violates Lorentz invariance. The inelasticity can be calculated by solving a third-order polynomial equation containing: a) the kinematics of the interaction, b) the LIV term for each particle and c) the geometry of the interaction. We use the inelasticity we calculate to investigate the proton propagation in the intergalactic media. The photopion production of the proton interaction with the CMB is taken into account using the inelasticity and the attenuation length in different LIV scenarios. We show how the allowed phase space for the photopion production changes when LIV is considered for the interaction. The calculations presented here are going to be extended in order to calculated the modified ultra-high energy cosmic rays spectrum and compare it to the data.

1. Introduction

Relativity is one of the most important theories of modern physics and one of its pillars is the Lorentz invariance, which is proposed to be a fundamental symmetry in Quantum Field Theory. The possibility of Lorentz invariance violation (LIV), however, has been investigated by several quantum gravity (QG) models and is incorporated in high energy models of space-time structure [1]. Testing LIV is, therefore, very relevant in order to investigate new proposals and also to set the range of validity of relativity.

In this work, we study the possibility of testing LIV with the propagation of ultra-high energy cosmic rays (UHECR). We use a generic first order perturbative term in the free particle Lagrangian [2], which results in a new momentum dependent term in the dispersion relation. This correction is expected to be suppressed by the Planck Scale. Thus, astrophysics is a good framework for testing it, since UHECR have been detected by the Pierre Auger Observatory [3] with energies three orders of magnitude larger than the events detected in particle accelerator.

Ultra-high energy cosmic rays interact with the low energy photon background as they propagate through the Universe. Protons interact mainly with the Cosmic Microwave Background (CMB) via pion and pair production [4, 5], heavier nuclei also interact with the Extragalactic Background Light (EBL), undergoing photo-disintegration [6] and energetic photons can produce an electron-positron pair when interacting with the CMB and the Radio Background (RB) [7]. The main structures in the UHECR spectrum could be caused by these interactions. The dip model proposes that in a pure proton spectrum, the ankle at $10^{18.7}$ eV appears naturally due to pair production [8]. On the other hand, one of the proposed models to explain the suppression at the highest energies could be due to pion production and photo-disintegration, which is called the GZK suppression [9, 10]. If LIV is considered, the energy losses of the UHECR in the interactions with the CMB are changed and, consequently, the energy spectrum structures too. For this reason, the UHECR spectrum, currently measured by the Pierre Auger Observatory [11] and by the Telescope Array experiment [12], can be a reliable source of information to impose limits in the LIV.

This paper is organized as follows: in Section 2 we present the LIV framework used. In Section 3 we discuss the modifications in the kinematics of interactions of propagating UHECR. In Section 4 we show the results in the photopion production. And, finally, in Section 5 we conclude the paper and discuss other important tests that could be studied.

2. Lorentz Invariance Violation

The Coleman and Glashow formalism, proposed in [2], is a formulation which has been used in several works to study cosmic rays propagation with LIV [13, 14, 15, 16]. It proposes a perturbation to the free particle Lagragian:

$$\mathcal{L} = \partial_\mu \psi^* Z \partial^\mu \psi - \psi^* M^2 \psi \tag{1}$$

$$\mathcal{L} \to \mathcal{L} + \partial_i \psi \delta_a \partial^i \psi \tag{2}$$

This new perturbative term must be calculated for each particles and it leads to a new dispersion relation given by:

$$E_a^2 = p_a^2 c^2 + m_a^2 c^4 + \delta_a p_a^2 c^2 \tag{3}$$

which can be rewritten as:

$$E_a^2 = p_a^2 c^2 (1 + \delta_a) + m_a^2 c^4 \tag{4}$$

It is then defined $c_a = c\sqrt{1+\delta_a}$:

$$E_a^2 = p_a^2 c_a^2 + m_a^2 \frac{c_a^4}{(1+\delta_a)^2} \approx p_a^2 c_a^2 + m_a^2 c_a^4 \tag{5}$$

therefore, c_a is the maximum attainable velocity of a particle or its "own speed of light".

The main effect of this perturbation is a shift in the center of mass system (cms) energy of a particle[1]:

$$\sqrt{s_a} = \sqrt{E_a^2 - p_a^2} = \sqrt{m_a^2 + \delta_a p_a^2} \tag{6}$$

The cms energy is crucial for the calculation of the energy threshold of interactions and, consequently, a shift in this quantity results in a shift in the threshold of such interactions.

[1] We were expressly showing c in the dispersion relation in order to compare with c_a, but from now on we will use natural units where $c = 1$.

3. Kinematics of Propagating UHECR

In order to calculate the effects of LIV in the energy spectrum of UHECR, it is necessary to calculate the energy losses of propagating cosmic rays considering LIV. In this work we present the calculation for any generic $a + b \rightarrow c + d$ interaction based on the calculations performed by Scully and Stecker [16].

Scully and Stecker assume the effects of LIV only in the inelasticity, leaving the cross section unchanged. The same assumptions are used in this work. Their calculations for the inelasticity, however, are performed in the nucleus reference frame (NRF) and, then, a boost to the laboratory frame[2] is necessary. Consequently, another assumption, that the boost is not modified by LIV is needed. The following calculations are already performed in the LF and, therefore, this assumption is avoided.

The inelasticity is obtained by imposing that the total rest energy, $\sqrt{s} = \sqrt{E_{tot}^2 - p_{tot}^2}$ is conserved, i.e., $s_i = s_f$. We can, then, obtain s_i and s_f in terms of the properties of the particles a, b, c and d:

$$\begin{cases} s_i = (E_a + E_b)^2 - (\vec{p}_a + \vec{p}_b)^2 = s_a + s_b + 2E_aE_b - 2p_ap_b\cos\theta_i \\ s_f = (E_c + E_d)^2 - (\vec{p}_c + \vec{p}_d)^2 = s_c + s_d + 2E_cE_d - 2p_cp_d\cos\theta_f \end{cases} \tag{7}$$

We then rewrite the momentum as a function of the energy and the mass and use that $E^2 >> s$:

$$s_i = s_a + s_b + 2E_aE_b\left(1 - \cos\theta_i + \cos\theta_i\left(\frac{s_a}{2E_a^2} + \frac{s_b}{2E_b^2}\right) + \mathcal{O}(2)\right) \tag{8}$$

$$s_f = s_c + s_d + 2E_cE_d\left(1 - \cos\theta_f + \cos\theta_f\left(\frac{s_c}{2E_c^2} + \frac{s_d}{2E_d^2}\right) + \mathcal{O}(2)\right) \tag{9}$$

In the regime where a is a relativistic particle interacting with a low energy background and generating two relativistic particles, which is valid for all the interactions of interest[3], the final angle can be approximated as $\cos\theta_f \approx 1$, therefore:

$$s_f \approx s_c + s_d + 2E_cE_d\left(\cos\theta_f\left(\frac{s_c}{2E_c^2} + \frac{s_d}{2E_d^2}\right)\right) \tag{10}$$

Finally, we want to look at the inelasticity of the interaction that is defined as the fraction of energy lost by the particle a due to the creation of c. In this regime, we can write:

$$K = \frac{E_c}{E_a} \implies E_c = KE_a \tag{11}$$

$$K = 1 - \frac{E_d}{E_a} \implies E_d = (1-K)E_a \tag{12}$$

Substituting E_c and E_d in $s_i = s_f$ and writing the cms energy of each particle using Eq. 6, we have:

$$\begin{aligned} &s_a + s_b + 2E_aE_b\left(1 - \cos\theta_i + \cos\theta_i\left(\tfrac{s_a}{2E_a} + \tfrac{s_b}{2E_b}\right)\right) \\ &= s_c(K) + s_d(K) + K(1-K)\left(\cos\theta_f\left(\tfrac{s_c(K)}{K} + \tfrac{s_d(K)}{(1-K)}\right)\right) \end{aligned} \tag{13}$$

[2] The laboratory frame (LF) is that in which the CMB is isotropic and is the frame in which the energy spectra of cosmic rays are usually presented.

[3] The pair production produces three particles: an electron, a positron and a photon. The electron-positron pair, however, can be approximated as a single particle with double the mass for the calculations.

It is highly desirable to write this expression as a polynomial, for it is easier to solve. Therefore, we first define:

$$A := s_i - m_c^2 - m_d^2 \tag{14}$$

and, then, multiply the expression[4] by $K(1-K)$:

$$\begin{aligned} -m_c^2 + K\left(A + 2m_c^2 - \delta_d E_a^2\right) + K^2\left(-A - m_c^2 - m_d^2 - \delta_c E_a^2\right) + K^3\left(\delta_c E_a^2\right) + \\ -\delta_d E_a^2 \sum_{i=0}^{3} K^{i+1}\binom{3}{i}(-1)^i - \delta_d E_a^2 \sum_{i=0}^{2} K^i \binom{2}{i}(-1)^{i+2} = 0 \end{aligned} \tag{15}$$

This third-order polynomial contains all the kinematics, LIV and geometry information and is easily solvable. Nevertheless, only roots in the regions $(0,1)$ are physical solutions for the problem.

Lastly, we use the modified inelasticity to calculate the attenuation length, ℓ, which is defined as the distance the cosmic ray can travel before losing $1/e$ of its energy:

$$\frac{1}{\ell(E)} = \frac{1}{E}\frac{dE}{dx} = \int_0^\infty d\epsilon n(\epsilon)\sigma(\epsilon) \int_{-1}^{1} dcos\left(\theta_i\right) K(\epsilon, E, \theta_i) \tag{16}$$

where $n(\epsilon)$ is the background density, $\sigma(\epsilon)$ is the cross section and the threshold appears naturally in the inelasticity, as $K = 0$ for $\epsilon < \epsilon_{th}$.

With this information it is possible to calculate the propagation for any cosmic ray including LIV effects. In the next section we show the results for protons producing pions ($p + \gamma \to p + \pi$) as an example.

4. Photopion Production

The photopion production with LIV has been widely studied [17, 14, 16, 18, 19, 20, 21, 22, 23] as this interaction modulates the shape of the spectrum in the highest energies, where the LIV effects could be strong enough to be detected. We have calculated the inelasticity and the attenuation length for this interaction using different values of δ_π and δ_p as an example for the calculations presented in the last section.

Figs. 1-3 show the inelasticity for the interaction as a function of the initial proton energy and the background photon energy, both in the LF. These figures also show the threshold energy for the background photon, which is the first energy with non-zero inelasticity, and the phase space of the interaction.

Turning on LIV for the pion reduces the phase space for the interaction as the threshold for the interaction becomes larger when compared to the LI case. This effect gets stronger as the initial proton energy gets larger. The opposite happens for the proton, the threshold is reduced, resulting in a larger phase space.

Fig. 4 and Fig. 5 show the attenuation length, i.e., the distance the proton can travel before losing $1/e$ of its energy, for different LIV terms. The changes in the phase space have direct effects in this. For the pion, the attenuation length gets larger and, therefore, the cosmic ray travels further before interacting. For the proton, on the other hand, it gets smaller, which means that the cosmic ray will interact more often.

The LIV term in the dispersion relation can be treated as an effective mass that gets larger with the energy. For the pion, a larger mass would mean more energy needed to create it,

[4] This is only possible as we know that $K \neq 0$ and $K \neq 1$. In the first situation there is no interaction and, in the second, only a pion is created, which is physically forbidden.

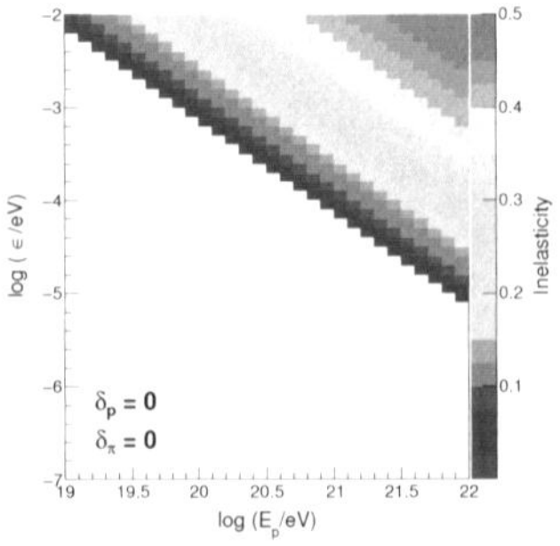

Figure 1. Inelasticity for the photopion production in a scenario without LIV.

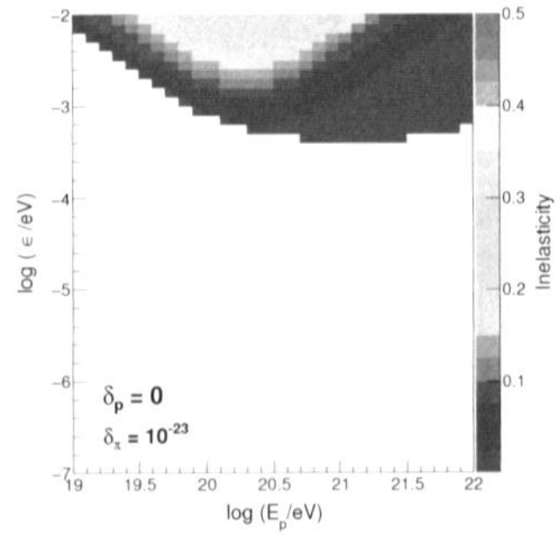

Figure 2. Inelasticity for the photopion production in a scenario with LIV for the pion.

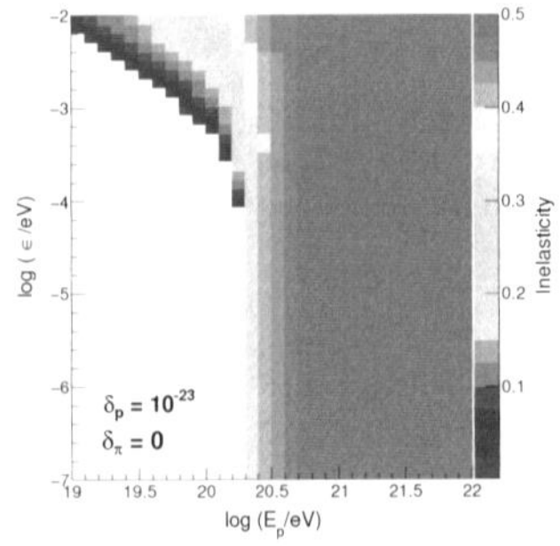

Figure 3. Inelasticity for the photopion production in a scenario with LIV for the proton.

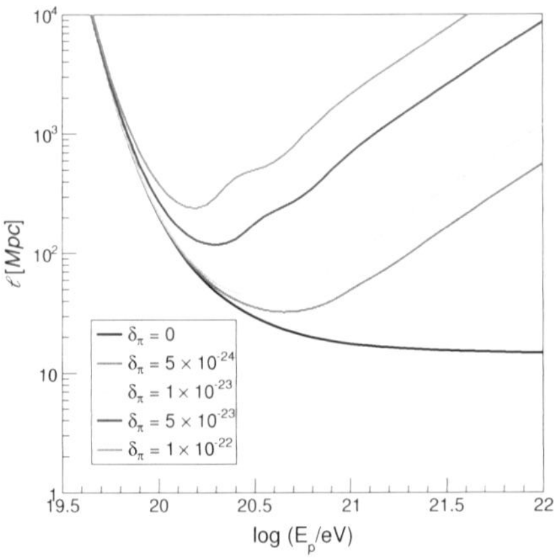

Figure 4. Attenuation length for the photopion production with LIV for the pion. Each line show the result for a different LIV term.

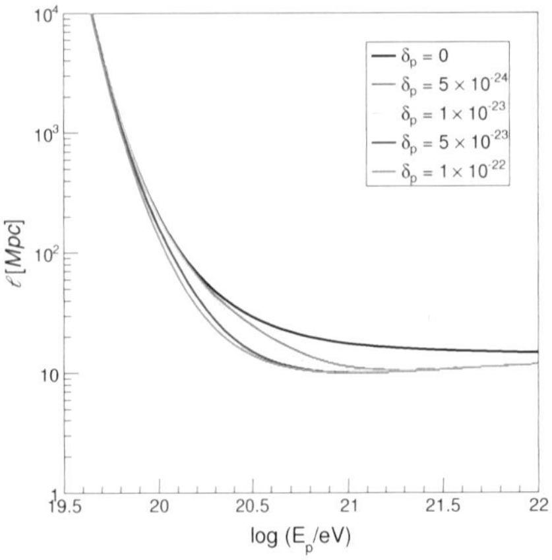

Figure 5. Attenuation length for the photopion production with LIV for the proton. Each line show the result for a different LIV term.

consequently shifting up the threshold. A shift in the proton mass, on the other hand, would change the energy both before and after the interaction. But, as the initial proton is more energetic, the shift before the interaction is larger and therefore the energy of the background photon necessary for creating the pion is smaller.

Those changes have direct influence in the UHECR spectrum as it could be recovered or suppressed in the highest energies.

5. Conclusions

The propagation of ultra energetic cosmic rays is a reliable source for testing Lorentz Invariance Violation as it leaves footprints in the energy spectrum measured with high statistics by different experiments.

In this work we have presented a generic analytic calculation for the inelasticity in the LF of any $a + b \rightarrow c + d$ interaction. The inelasticity is directly related to the threshold of these interactions and is used to obtain their energy losses. We have shown, as an example, the results of the calculations for the photopion production. The main effect is a change in the phase space of the interaction, which can be larger or smaller, depending on the LIV terms.

The calculations here proposed can be used in several LIV studies. Limits for the LIV terms, for example, could be imposed by performing these calculations for all the interactions present in the propagation and obtaining the spectrum using propagation codes such as CRPropa [24] and SimProp [25].

Acknowledgments
We would like to thank the São Paulo Research Foundation (FAPESP) for the financial support through Grant No. 2014/26816-0 and 2010/07359-6 and Conacyt for the support through the Mesoamerican Center for Theoretical Physics, provinding a scholarship to Rodrigo Guedes Lang in order to participate in the school and present this work. We would also like to thank Humberto Martinez for the help provided to fully understand and double check the calculations.

References

[1] Mattingly D 2005 *Living Reviews in Relativity* **8** 1–78
[2] Coleman S and Glashow S L 1999 *Physical Review D* **59** 116008
[3] The Pierre Auger Collaboration 2015 *Nuclear Instruments and Methods in Physics Research, Section A: Accelerators, Spectrometers, Detectors and Associated Equipment* **798** 172–213
[4] Stecker F W 1968 *Physical Review Letters* **21** 1016–1018
[5] Blumenthal G R 1970 *Physical Review D* **1** 1596–1602
[6] Stecker F W and Salamon M H 1999 *The Astrophysical Journal* **20** 521–526
[7] De Angelis A, Galanti G and Roncadelli M 2013 *Monthly Notices of the Royal Astronomical Society* **432** 3245–3249
[8] Berezinsky V 2009 *Nuclear Physics B - Proceedings Supplements* **188** 227–232
[9] Greisen K 1966 *Physical Review Letters* **16** 748–750
[10] Zatsepin G T and Kuzmin V A 1966 *Zh. Eks. Theor. Fiz. Pis'ma Red* **144**
[11] Inés Valiño for the Pierre Auger Collaboration 2015 *Proceedings of the 34th International Cosmic Ray Conference* (The Hague, The Netherlands)
[12] Abu-Zayyad T et al 2015 *Astroparticle Physics* **61** 93–101
[13] Jacobson T, Liberati S and Mattingly D 2002 *Physical Review D* **67** 35
[14] Stecker F and Scully S 2005 *Astroparticle Physics* **23** 203–209
[15] Maccione L, Taylor A M, Mattingly D M and Liberati S 2009 *Journal of Cosmology and Astroparticle Physics* **2009** 022–022
[16] Scully S and Stecker F 2009 *Astroparticle Physics* **31** 220–225
[17] Alfaro J and Palma G 2003 *Physical Review D* **67** 083003
[18] Bi X J, Cao Z, Li Y and Yuan Q 2009 *Physical Review D* **79** 083015
[19] Bietenholz W 2011 *Physics Reports* **505** 145–185
[20] Saveliev A, Maccione L and Sigl G 2011 *Journal of Cosmology and Astroparticle Physics* **2011** 046–046
[21] Cowsik R, Madziwa-Nussinov T, Nussinov S and Sarkar U 2012 *Physical Review D* **86** 045024
[22] Boncioli D, di Matteo A, Salamida F, Aloisio R, Blasi P, Ghia P L, Grillo a F, Petrera S and Pierog T 2015 *Proceedings of the 34th International Cosmic Ray Conference* (The Hague, The Netherlands) pp 1–8
[23] Aloisio R, Blasi P, Ghia P L and Grillo A F 2000 *Physical Review D* **62** 053010
[24] Batista R A, Erdmann M, Evoli C, Kampert K H, Kuempel D, Mueller G, Sigl G, Van Vliet A, Walz D and Winchen T 2015 *Journal of Physics: Conference Series* **608** 012076
[25] Aloisio R, Boncioli D, Grillo A, Petrera S and Salamida F 2012 *Journal of Cosmology and Astroparticle Physics* **2012** 007–007

Detection and analysis of astroparticles using WCD at 2800 m a.s.l. in Quito

A Correa[1] S Vargas[1] N Vásquez[1] and A Delgado[1], for the LAGO Collaboration[2]

[1] Escuela Politécnica Nacional, Department of Physics, Ladrón de Guevara E11253, Quito-Ecuador
[2] http://lagoproject.org, see the full list of members and institutions at lagoproject.org/collab.html (lago-pi@lagoproject.org)

E-mail: melida.correa@epn.edu.ec

Abstract. At the Escuela Politécnica Nacional we have assembled a WCD (Water Cherenkov Detector) prototype for the LAGO (Latin American Giant Observatory) project in Ecuador. This article presents the data as well as the analysis corresponding to October, 2015. We present the obtained Charge Distribution Histogram (CDH). We shaped the conditions in which the equipment is operating given the environmental parameters and the value for the first VEM (Vertical Equivalent Muon) for the "Politanque".

1. Introduction

The Latin American Giant Observatory (LAGO) project is a collaboration created in 2005 which consists of 10 countries, with more than 80 scientists and students of different disciplines [1]. The detection of the high energy component in Cosmic Rays (CR) takes advantage of places located above 4000 meters above the sea level. The main objectives of this project are: study the high energy component of CR, transitory events, long term events, climate and space weather. We can also study solar modulation, monitor CR and do other analysis.

Fig. 1 presents three detectors in Ecuador that are located in: ESPOCH at 2750 m a.s.l. - Chimborazo province, USFQ and EPN at 2800 m a.s.l. - Pichincha province.
Astroparticle Laboratory of Escuela Politécnica Nacional (EPN) is where the third detector- "Politanque" is located, it works with undergraduate and master students, and scientists to encourage the study of astroparticles in different fields, from construction of a detector to the study of the theory that this implies.

1.1. Cosmic Rays

The energy spectra for individual elements of cosmic rays are obtained directly with satellite and balloon experiments at the top of the atmosphere; these experiments can detect CR with energies up to 10^{15} eV. Due to the fast decreasing flux, measurements at higher energies require large detection areas or long exposure times, which can only be obtained in groundbased detector systems. These detectors measure extensive air showers generated by interactions of the

IOP Conf. Series: Journal of Physics: Conf. Series **866** (2017) 012009 doi:10.1088/1742-6596/866/1/012009

Figure 1. Political map of Ecuador showing LAGO locations. The Chimborazo province where the "Chimbito" detector is located is printed in red, and the Pichincha Province where the "Panchito" and "Politanque" detectors are located, in blue. [3]

high energetic cosmic rays with nuclei in the atmosphere [2].

Cosmic Rays (about 90 % protons, 9 % alpha particles and other heavy nuclei) hit the Earth's atmosphere where they ionize the atoms. Most relativistic cosmic rays have energies comparable to or greater than its rest energy. Few of them have ultra-relativistic energies exceeding 10^{20}eV, eleven orders of magnitud larger than the rest energy of the proton [4].

The particles that come from the Sun are relatively few and are characterized by temporal association with violent events in the sun and thus by a rapid viariability [4]. In contrast, cosmic rays are anticorrelated with solar activity, therefore during the increased periods of solar activity the CR flux on the Earth is lower. Cosmic rays with very high energy have typical Larmor radii which are larger than the size of the galaxy which points to their extragalactic origin.

The primary cosmic rays hitting the outer layers of the atmosphere undergo collisions with atmospheric nuclei. These collisions result in showers of new elementary particles of all kinds (such as electrons, positrons, pi mesons, muons, etc.). These showers eventually reach the surface. The secondary cosmic ray showers may reach an area of several square kilometers.

To estimate the number of possible interactions of the particles (generated in the showers) traversing the atmosphere, we need to first determine the mean free path, [2, 4]

$$\lambda_i = \frac{N_A}{A_i}\sigma_i[\frac{gr}{cm}]. \tag{1}$$

Where N_A is Avogadro number, A atomic number, σ_i cross section of interaction - Atmospheric depth. The indirect detection of cosmic rays depends on the energy of the primary particle which arrives the top of Earth's atmosphere, the angle of incidence, the magnetic rigidity and height of the first interaction.

Considering the proton as the primary particle that arrives perpendicularly to the location of Ecuador, the number of muons generated in the showers increases with the depth of the atmosphere because the interaction cross-section for the incident proton increases with atmosphere depth. The increasing interaction cross-section is illustrated by the number of generated muons with atmospheric depth in Figure 2.

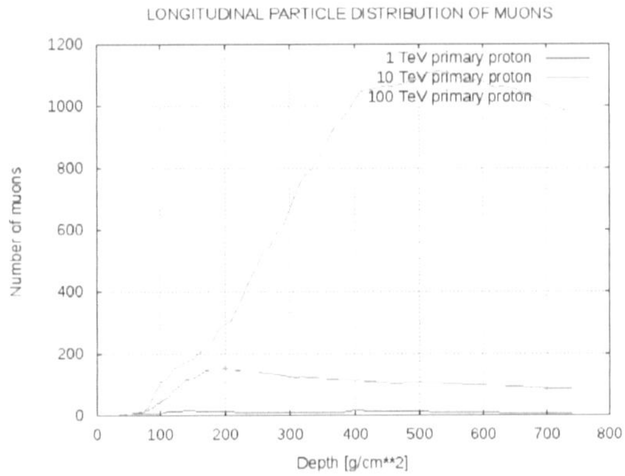

Figure 2. Longitudinal Distribution of Muons from a primary proton with different energies 1 TeV, 10 TeV and 100 TeV.

Muons and neutrinos are the most penetrating components of the secondaries cosmic rays, they are the result from the interaction mechanisms that produce pions and kaons through the decay chains. Pions(π^+, π^-) decay into muons (μ^+, μ^-) and neutrinos ($\nu_\mu, \bar{\nu_\mu}$). Since muon is a lepton, it decays to an electron or a positron plus neutrinos.

$$\mu^- \to e^- + \bar{\nu_e} + \nu_\mu \; ,$$
$$\mu^+ \to e^+ + \nu_e + \bar{\nu_\mu} \; .$$

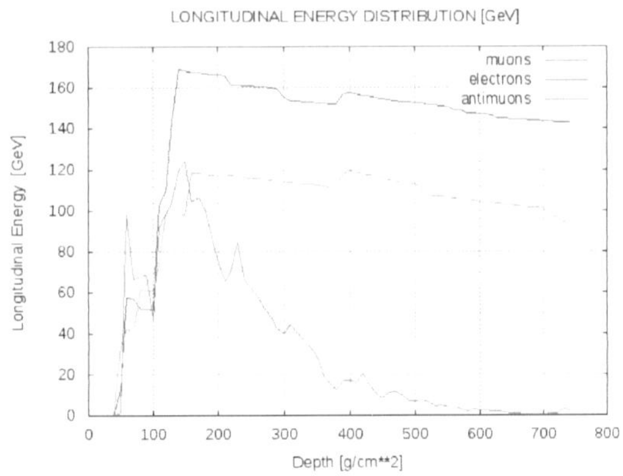

Figure 3. Longitudinal Energy Distribution of muons, antimuons and electrons from a primary 1 TeV proton.

In Fig. 3 we plot the energy of selected interaction products (generated by a 1 TeV proton) as they traverse the atmosphere. Muons typically lose 2 GeV while transiting the atmosphere due to ionization, consequently, muons dominate the charged-particle spectrum at the earth's surface (they also lose energy by bremsstrahlung and production of e^- e^+ pairs, photonuclear reactions like Compton production, etc.)

1.2. Cherenkov Effect

A charged particle that moves through a medium where the atoms forming the material are arranged and connected to each other, in a moment of time, in the region closer the path of the charged particle, the electric field of the particle distorts atoms, moving their charge causing polarization around that point. If the particle moves faster than the speed of light in the azimuthal plane, the symmetry is preserved, but along the axis there is a resultant of the dipole field and the bias field is not symmetrical, this field is momently created to the particle by a short electromagnetic pulse. [8]

Taking the principle of Huygens, the waves radiated that are in phase with each other resulting in a field at a distance from the observation point. This radiation is observed at an angle θ with respect to the trajectory of the particle, these waves are coherent and are combined in a plane wave. [9] As the particle travels faster than the speed of light in the medium, it produces Cherenkov radiation and it is given by:

$$cos\theta = \frac{1}{n\beta}. \tag{2}$$

Where θ is the emission angle, n is the refractive index of the medium and $\beta = \frac{v}{c}$ is the relativistic velocity coefficient.
In general:

$$\frac{d^2W}{dw.dx} = \frac{e^2w}{c^2}\left[1 - \frac{c}{n\nu}\right]w. \tag{3}$$

The fundamental features depend only on the charge and not of the mass of the moving particles, as well as the spectral dependency on the frequency is linear for the homogeneous medium.[8, 9]
This effect is used to determine muons which are the most energetic particles when they arrive at the earth's surface, and they are part of the secondary particle showers where the number of particles generated is directly proportional to the energy of the primary particle.[8]

Single particle technique is to take a count of these particles to estimate the energy of the generating particle. This count depends on the height above sea level where the detector is located and the mean-life time of the secondary particles [5].

2. Experimental setup

The "Politanque" detector is installed at the Escuela Politécnica Nacional (EPN). It's a small-sized detector accessible to students and teachers from within the university as well as other collaborating scientists. Its aim is to serve as a developing detector set up. "Politanque" has been fully operational for a small period of time. [3]

2.1. WCD "Politanque" and Data acquisition

The detector consists of a black polyethylene tank (volume $0.11\,\mathrm{m}^3$, height $0.23\,\mathrm{m}$ and radius $0.23\,\mathrm{m}$); a 5" EMI 9530A photomultiplier, a high-voltage source and a data acquisition system. The hardware and software used for data acquisition are from the EAS-BUAP project [3].
This system works with a voltage input $\sim 1450[V]$ for the PMT where the output signal is collected in a digitizer board DAQmx (based on Dual ADC AD9216-100MS/s, 10bit resolution, 4-channel input), the information is then processed by a Nexys2 Board programed using VHDL, for being stored and processed in a PC with Ubuntu 12.04 (See Fig. 5).

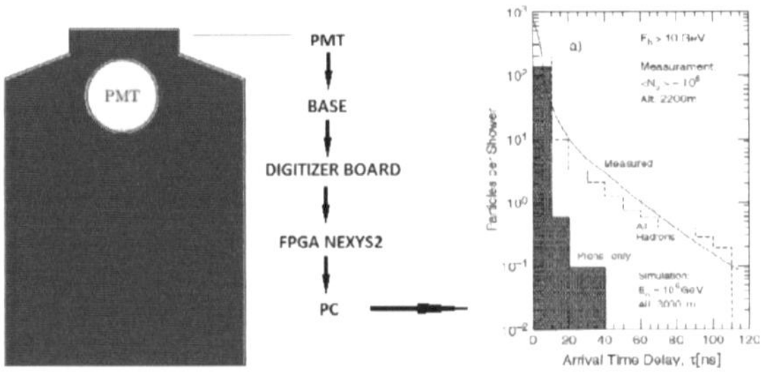

Figure 4. Experimental setup scheme.

Figure 5. Experimental setup; the electronics used for the experiment are from the EAS-BUAP project.

The USB-Serial Port connection from the preconfigured Nexys2 Board to the computer allows us to manipulate, via VHDL programing, the thresholds of the sampled data. The acquired data is in ADC in the range of nanoseconds. Each pulse has 16 samples of 10 nanoseconds each one [7].

3. Data analysis

The main characteristic of the detection of cosmic rays is the amount of energy deposited by each particle from 10 GeV to 100 GeV.

The script file adds a column of charge (x-axis), generating the *nuevos.dat* file; *picos_discriminados* counts the peaks based on a reasonable minimum height level (counts) and the deviation (to be defined). Fig. 6 shows 7 of such peaks obtained from a measurement.

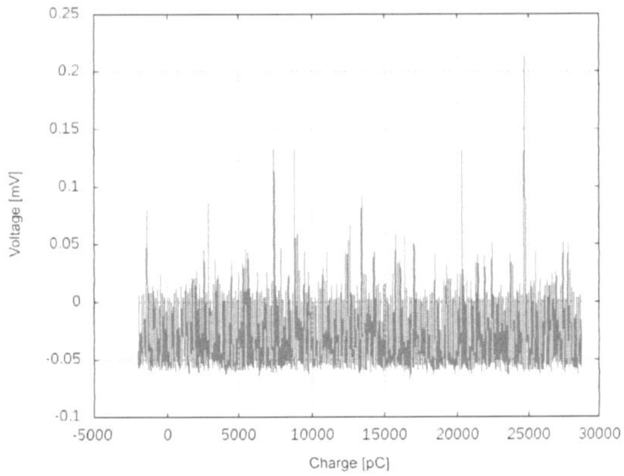

Figure 6. Diagram for counting peaks that work to VEM

Distribution of integrated charge of particles is depicted in the histogram in Fig. 7. Muons entering through the top and exiting through the side, muons entering and exiting through the side, and small showers that produce the first peak, as well as γ and e^- .

IOP Conf. Series: Journal of Physics: Conf. Series **866** (2017) 012009 doi:10.1088/1742-6596/866/1/012009

The main analysis is done with the Vertical Equivalent Muon (VEM) calibration that describes the average charge of the tank when it is crossed by a single incident vertical muon on the center of the top cover [6, 10].

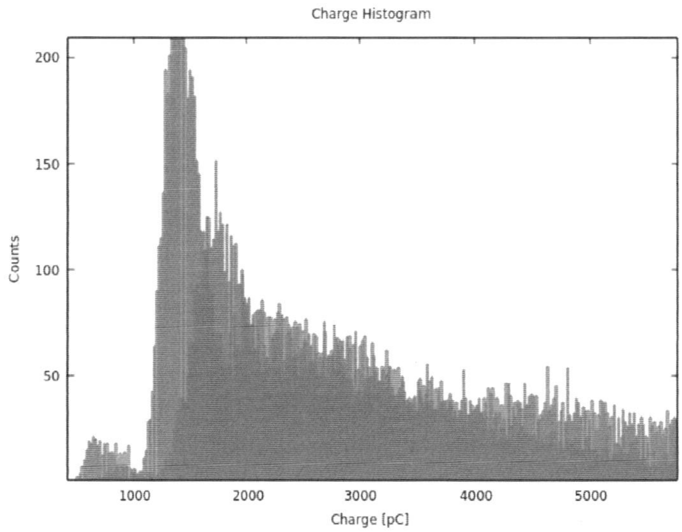

Figure 7. Charge Distribution Histogram (CDH) for "Politanque" detector

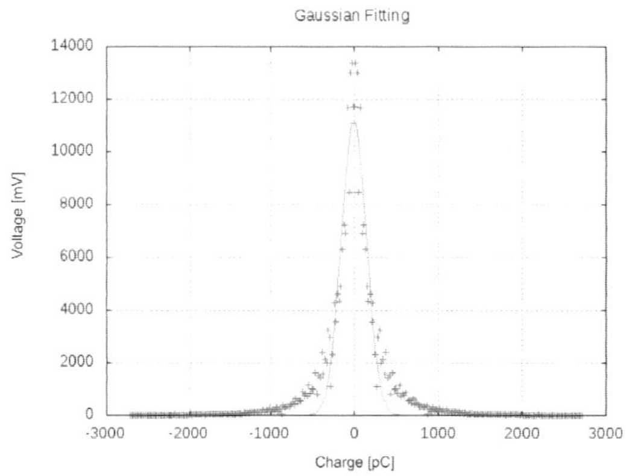

Figure 8. Vertical Equivalent Muon(VEM) calibration adjusted by a Gaussian fit. We found a value of 499.000 pC with a statistical error of 106.

The results obtained from the analysis of Charge Distribution Histogram (CHD) for a sample of 770000 events are presented in the Fig. 7. The peak value of 499.000 ± 101 pC for the VEM was determined by a Gaussian fit presented in Fig. 8.

With this we can also see the quality of water, because, if the water is pure enough, muons will produce Cherenkov effect and photons will be able to lose a minimum of absorption energy, reaching the PMT after several bounces and achieving detection.

4. Conclusions

The Data obtained with "Politanque" show that for a better fit of the VEM, the acquisition periods should last longer. Additionally, following the special behavior of the CR energy spectrum, a high energy threshold demands a higher energetic event, which was found in the relation between the number of events and the thresholds at the same exposure time. We took into account that higher energetic events have the property of decrease its frequency.
Cosmic Rays particles that reach the Earth atmosphere produce new particles. In their propagation through the atmosphere they can interact with the molecules in the air or they can decay. In the performed simulations with proton as primary particle, a shower of new particles is obtained by the proton interaction with the atmosphere. As the shower increases its atmospheric depth, muons predominate as the most energetic particles, which can be seen in the longitudinal energy distribution of the particles.

5. Acknowledgements:

The LAGO Collaboration is very thankful to the Pierre Auger Collaboration for its continuous support. The authors of this paper gratefully acknowledge the financial support from Escuela Politcnica Nacional, Universidad San Francisco de Quito USFQ, Fundacin CEDIA through CEPRA IX research grants and SENECYT for the prometeo program. And I am grateful to Conacyt for the support trough the Mesoamerican Center for Theoretical Physics.

6. References

[1] The Latin American Giant Observatory (LAGO), (2014). Retrieved Sep. 23, 2016, from http://lagoproject.org/
[2] Grieder, P. K. (2010). *Extensive Air Showers (p. 303).* Heidelberg: Springer.
[3] Vargas, S., Mantilla, C., Martinez, O., & Vásquez, N. for the LAGO collaboration. (2015) *Implementing a set of WCD detectors for Space Weather research: first results and further developments.*, Proceedings, 34rd International Cosmic Ray Conference (ICRC2015).
[4] Gaisser, T., Protheroe, R., Turver, K., & McComb, T. (1978). *Cosmic ray showers and particle physics at energies 10 15-10 18 eV.* Reviews of Modern Physics, 50(4), 859.
[5] Asorey, H., Dasso, S., Nez, L., Prez, Y., Sarmiento-Cano, C., & Surez-Durn, M. (2015). *The LAGO space weather program: Directional geomagnetic effects, background fluence calculations and multi-spectral data analysis.* The 34th International Cosmic Ray Conference, Vol. PoS (ICRC2015).
[6] Cotzomi, J., Martínez, O., Moreno, E., Salazar, H., Villaseor, L. (2005). *Air shower array at the university of Puebla for the study of cosmic rays,* Revista Mexicana de Física, 51(1), 38-46.
[7] Sánchez, C., Rubén, J. (2015). *Instrumentación para el Arreglo de Detectores de Superficie LAGO-México,* Benemérita Universidad Autónoma De Puebla, PHD Thesis, México.
[8] Grieder, P. K. (2010). *Primary Cosmic Radiation and Astrophysical Aspects.* Extensive Air Showers, 479-588.
[9] *Gaisser, T. K., Engel, R., & Resconi, E. (2016, June 2). Cosmic rays and particle physics.* Cambridge University Press.
[10] Asorey H. for the LAGO Collaboration. (2015). The LAGO Solar Project, Proceedings, 34th International Cosmic Ray Conference (ICRC2013), Rio de Janeiro, Brazil.

IOP Conf. Series: Journal of Physics: Conf. Series **866** (2017) 012010 doi:10.1088/1742-6596/866/1/012010

Cosmic ray muon study with the NEVOD-DECOR experiment

Oscar Saavedra San Martin

Istituto Nazionale di Astrofisica, INAF, Italy

saavedra@to.infn.it

Abstract. The experiment NEVOV-DECOR, which is desinged to study the cosmic muons at very inclined directions, is running under the collaboration of the Moscow Engineering Physics Institute, Moscow, Russia, and the Instituto Nazionale di Astrofisica and the Dipartimento di Fisica, Università di Torino, Italy. The main purpose of this experiment is to study the characteristics of the high multiplicity muons in muon bundles and their angular distributions. The result has shown the observation of the second knee at 10^{17} eV in the primary cosmic ray spectrum. In addition, we found that the number of high energy muons in EAS is more than 30% of what is predicted by the Monte Carlo models. This effect was found also by other experiments like Auger, but at primary cosmic ray energies higher than 10^{18} eV. We will present and discuss the main results of these investigations.

1. Introduction

To resolve the problem of the origin, acceleration and propagation of high energy cosmic rays, we need to know their energy spectrum and mass composition. If the energy of the primary cosmic ray is higher than 10^{15} eV, the only way to know the characteristics of the cosmic ray particles is through the study of Extensive Air Showers (EAS) at ground level. They are produced when the primary cosmic ray interacts high in the atmosphere.

The EAS are studied by means of their components: the electromagnetic component (with photons, electrons and positrons), the hadronic component (with protons, neutrons, pions, etc.), the leptonic component (with muons and neutrinos), and other signals produced by charged particles like Cherenkov light in the atmosphere, luminescence light, radio waves etc.

To detect each component, many types of detectors are used, like scintillators, water Cherenkov detectors, etc. They are located in many laboratories distributed in several places, and altitudes above sea level, or in underground laboratories, according to the purpose of the experiment and which of the component is being studied. The arrival directions of cosmic rays are isotropic, however if we study their zenith angular distribution there appears an absorption effect on the observed EAS due to the thickness of the atmosphere, from observations of the charged secondary particles in the EAS, and even more due to the Earth magnetic field.

In the primary interaction many secondary particles are produced, in particular, π's. Such particles can decay into μ's. The μ's suffer a feeble interaction through the atmosphere, therefore, they can transport direct information from the first hadronic interactions and consequently information about the primary particle. It is very important to study the cosmic muons mainly at very large zenith angle since most of the other components are absorbed in the atmosphere and the only survival component is the muonic one. The main objective of the cooperation between MEPhI and Torino institutions was the construction of the large area coordinate-tracking detector DECOR around the Cherenkov water calorimeter NEVOD [1] for investigations of multiple particle cosmic ray events at large zenith angles. The description of basic results of multi-muon event analysis are presented in this document. In particular, two new methods are described. They are connected with the EAS muon component investigation: the local muon density, D_μ (which is realized in the NEVOD-DECOR experiment [2]) and the muon bundle energy deposit, E_μ.

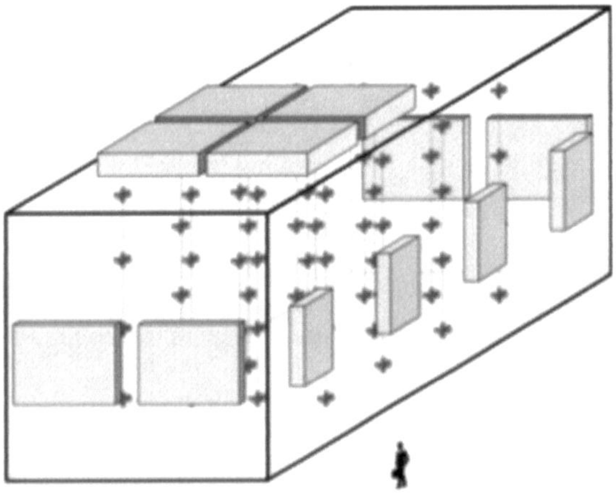

Figure 1. General layout of the NEVOD-DECOR experiment [2] (see the text for a description of the main components).

2. Experimental data

The general layout of the NEVOD-DECOR experimental complex is shown in fig 1. The NEVOD detector consists of a water Cherenkov tank with 2000 m^3 and is formed by a spatial lattice of quasi-spherical (QSM) optical modules each of them consisting of six photomultipliers with flat 15 cm diameter photocatodes. Thus the detecting latice has a nearly cubic shape, in total it includes 91 QSM. The side part of the coordinate-tracking detector DECOR is deployed in the galleries of the NEVOD building from three sides of the water tank. Two pairs of supermodules (SM) are located in the opposite short galleries and four supermodules in the long gallery. Each SM has an effective area of 8.4 m^2 and consists of eights planes of streamer tube chambers with resistive cathode coating hanged vertically with 6 cm spacing. The chambers are operated in a limited streamer mode with a continuous gas flow of an Ar+CO2+n-pentane mixture. The length of each chamber is 3.5 m. The planes of the chambers are equipped with a two-dimensional system of external read out strips (256 X- and 256Y-channels in each plane with 1.0 and 1.2 cm pitch, respectively). As a first level trigger condition for the SM, a 4-fold coincidence of signals from at least 2 odd and 2 even planes is used. The accuracy of the reconstruction of the tracks crossing the SM is about 1 cm in space and better than 0.7° and 0.8° for the projected zenith and azimuth angles, respectively.

In 2002-2007 a long series of measurements with the NEVOD-DECOR experiment were conducted. Net operation time with good quality data for the full configuration of the side DECOR was 19,922 h. The basic set of event selection conditions were based on the coincidences of the signals between different parts of the coordinate detector. The signals of the NEVOD QSM lattice and the CTS (a calibration telescope system that consists of two planes of scintillation detectors) were kept unchanged during the measurements. Selection of inclined muons bundles from DECOR data is based on the fact that tracks of muons originated in the atmosphere are nearly parallel. The data are analyzed part by part, for certain ranges of multiplicities and zenith angles. Phenomenological distributions of muon bundles as a function of particle multiplicity for several zenith angle intervals and in zenith angle for three values of the minimal number of muons are presented in figure 2. The obtained data on muons cover 6 orders of magnitude in the event intensity.

3. Local muon density spectra (LMDS): results and discussions

At zenith angles larger than 50-70°, depending of the primary particle energy, EAS are detected at the ground level as, practically, muon component. Due to a large distance from the EAS generation point to the observation level, the transversal dimension of shower muons rapidly increases with the zenith angle, the effect being additionally enhanced by the particle deflection in the geomagnetic field. Therefore, a muon detector with sizes of tens of meters may be considered as a point-like probe. In an individual event, the local density of EAS muons at the observation point may be estimated as the ratio of the muon bundle multiplicity m to detector area for a given direction: $D_{\mu}{\sim}m/S_{\mathrm{det}}$. [4,5].

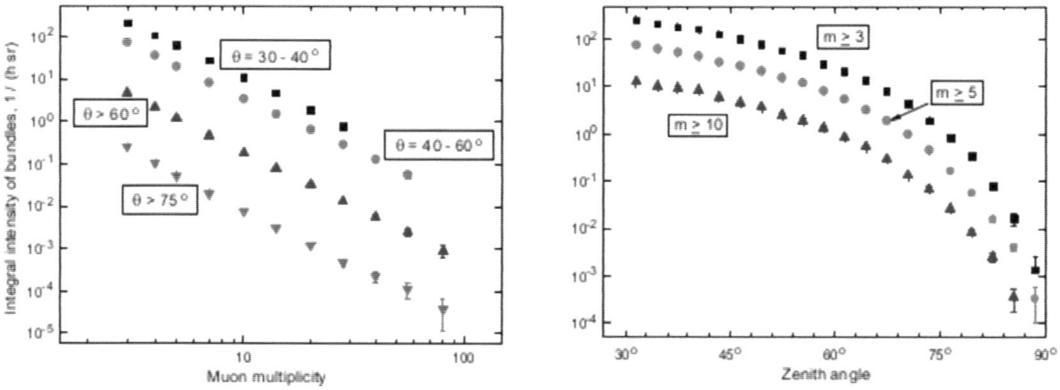

Figure2. Integral distribution of muon bundles as a function of particle multiplicity (m) for several ranges of zenith angle (left) and as a function of the zenith angle distribution of the bundles for different minimal number of muons.

Primary particles with different energies contribute to events with a certain muon density at distinct distances from the axis. However, due to the fast decrease of the primary cosmic ray (PCR) intensity, at increasing energies, the effective primary energy band appears relatively narrow (see figure 3, left).

It is important to note that at different zenith angles events with a fixed muon density are created by primary particles with substantially different energies that is, the event collection area is determined not by the detector size, but by the dimensions muon component in the EAS which near the horizon reach several kilometers. This fact allows to explore a very wide range of PCR energies by means of the LMDS method in a single experiment with a relative small detector. The dependence of the effective EAS collection area on zenith angle is shown in figure 3, right. At large zenith angle an important role is played by the Earth magnetic field (EMF) and the S_{eff} is sensitive to the field strength and direction. For zenith angles larger than 80 °, S_{eff} exceeds 1 km^2, which is sufficient to collect acceptable event statistics at energies near 10^{18} eV [4].

The distribution of the events in muon density-local muon density spectra- is sensitive to the shape of the spectrum of primary cosmic rays. Similarly to the spectrum of EAS in the number of muons, LMDS vary (mainly in absolute intensity) with the primary composition. Local muon density spectra are formed mainly by the central part of the EAS and are determined by the most energetic muons, and the most energetic parent hadrons, propagating near the shower axis, hence they carry additional information about the forward kinematical region of hadron interactions. A comprehensive description of the phenomenology of the LMDS technique and features of its application to the analysis of DECOR data on muon bundles was recently published in [3].

A general approach in the analysis of muon bundle data includes the following steps: reconstruction

IOP Conf. Series: Journal of Physics: Conf. Series **866** (2017) 012010 doi:10.1088/1742-6596/866/1/012010

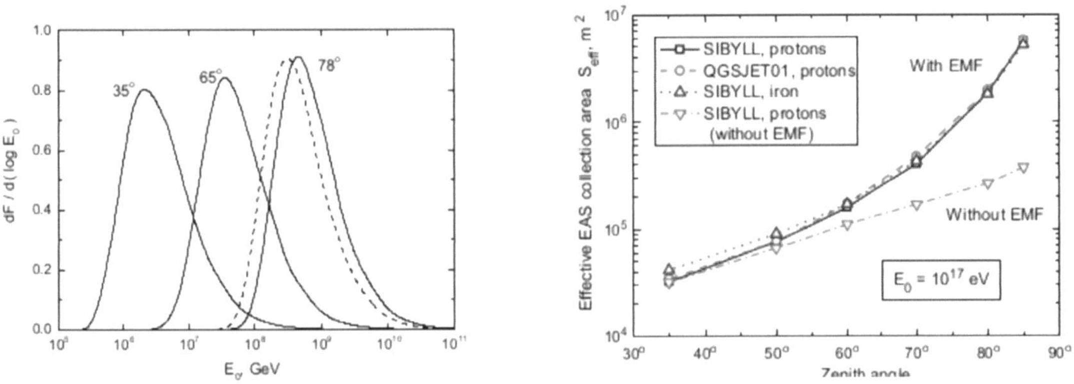

Figure 3 Left: Distribution of primary cosmic ray particle energies contributing to events with a fixed muon density (D,=0.2 muon/m^2) at different zenith angles. Solid curves correspond to calculations for primary protons, dashed curves, for iron nuclei. Right: Dependence of the effective EAS collection area on zenith angle in the LMDS technique.

of local muon density spectra in a detector-independent form from the measured distribution of muon bundle characteristics, taking into account Poisson fluctuations of the number of muons that hit the setup, triggering and selection conditions, detector efficiency, etc.; calculation of the expected LMDS for various assumptions about the primary spectrum, mass composition and hadron interaction models on the basis of 2-dimensional muon lateral distribution functions (LDF) simulated by means of the CORSIKA [5] code taking into account the geomagnetic field effects; and comparison of data with calculated results.

In figure 4, experimental LMDS reconstructed from DECOR data on muon bundles are compared with CORSIKA-based calculations for 4 different zenith angles (35°,50°, 65° and 78°). For comparison, a model of the primary flux with a power-law energy spectrum with differential slope +2.7 below the knee energy (4 PeV) and steepening to -3.1 above the knee was assumed. Absolute normalization was chosen according to the bulk of the experimental data around the knee presented in the review [6], two limiting cases of the mass composition (only protons and only iron nuclei) were considered. The curves in the figure correspond to two widely used hadron interaction models available in CORSIKA: SIBYLL 2.1 and QGSJET01 (dashed and solid curves, respectively). Arrows in the upper part of each frame indicate the values of effective (mean logarithmic) primary energies.

As it is seen from the results for moderate zenith angles (θ=35°, upper left frame) at energies around the knee the experimental LDMS is in a reasonable agreement with calculations, including the absolute normalization. The change of the LMDS slope is related with the knee in the primary spectrum. These results may be considered as an indirect check of the energy scale calibration.

At effective primary energies $E_0 > 10^{16}$ eV and zenith angles equal to 50° and 65° a progressive increase of the measured LMDS in comparison with calculations is observed, which can be interpreted as a trend to a heavier mass composition, though an alternative interpretation related with increasing deficit of muons in simulated EAS is also possible. Near primary energies equal to 10^{17} eV, a hint for a second knee was found [7], partial fits of the data obtained at zenith angle equal to 65° below and above this energy indicate the increase of the integral LMDS slope, β, by 0.20 ± 0.09. This result is also in quantitative agreement with slope estimates derived from the data at other zenith angles (thin lines in the figures, the intervals of $E_o = 10^{16}$ -10^{17} for 50° and $E_o > 10^{17}$ eV for 78°). By combining statistically independent data for different angles, we obtain Δβ = 0.15 ± 0.04. Taking into account the relation between the LMDS slope β and the integral primary energy spectrum index γ, i.e. $\beta \sim \gamma/0.9$ [4], it corresponds to an increase of the primary spectrum slope by $\Delta\gamma = 0.13 \pm 0.04$.

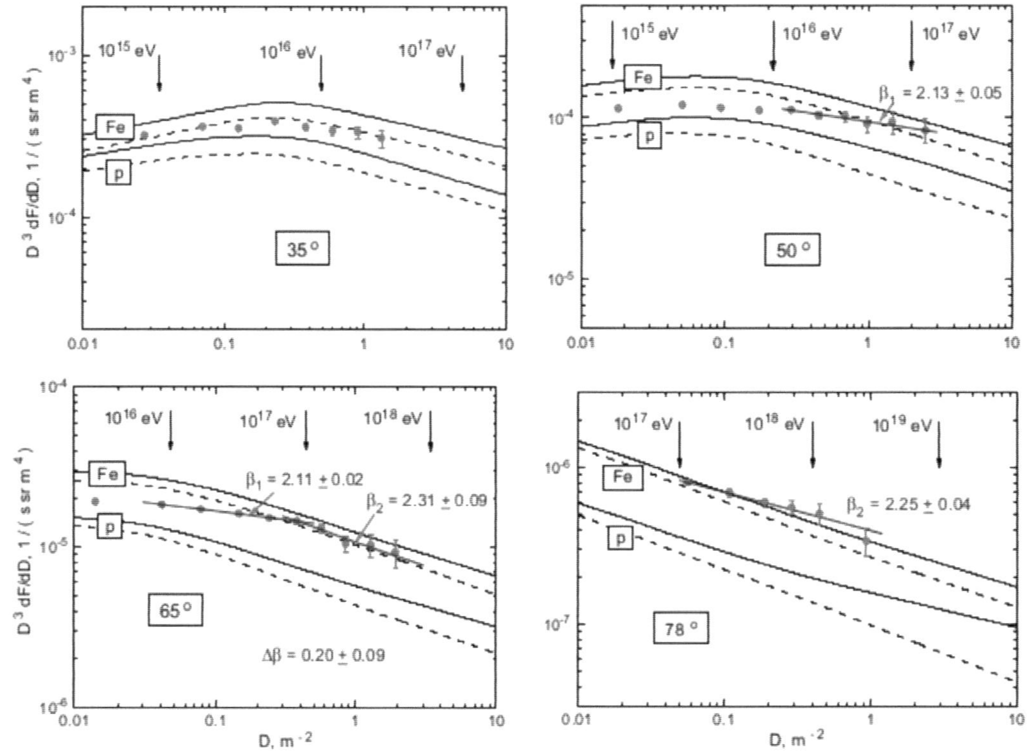

Figure 4. Measured (points) and calculated differential local muon density spectra for 4 zenith angles (labels are shown in the frames). Thin lines represent partial power-law fits of the data points between 10^{16} and 10^{17} eV (integral spectrum slope β_1) and above 10^{17} primary energy (integral spectrum slope β_2).

The conclusion about the steepening of the all-particle cosmic ray spectrum near 10^{17} eV was confirmed by KASCADE-Grande experiment [8].

Finally, data points at high muon densities and the largest zenith angle (78°), which correspond to primary energies around 10^{18} eV, lie near the upper edge of the calculation uncertainty band and even somewhat exceed the expectation for pure iron flux.

The relation between the local muon density spectrum and the primary cosmic ray flux is determined by the muon LDF, which in turn depends on the hadron interaction model and primary particle type. Therefore under certain assumptions about primary mass composition and hadron interaction model, the LMDS may be converted into an estimation of the primary cosmic ray spectrum [9]. Such procedure was applied to data samples with high muon multiplicities ($m \geq 10$) and two zenith angle intervals with: $\theta \geq 80°$. These data correspond to the mean logarithmic primary energies around 10^{18} eV and higher. The conversion was performed with two assumptions on the primary composition (pure protons and pure iron nuclei) using five different hadron interaction models available in the recent CORSIKA versions: SYBILL 2.1, QGSJET01, QGSJET-II, EPOS1.61 and EPOS1.99. For every sample 10 estimates of the differential primary energy spectrum has been obtained.

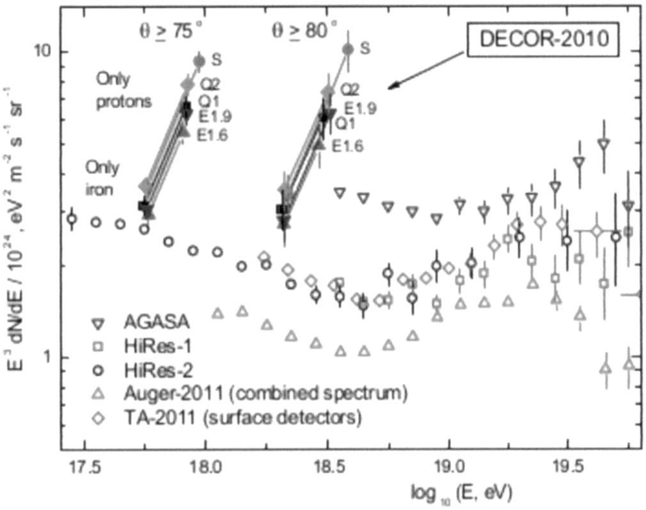

Figure 5. Comparison of differential primary spectrum estimates derived from DECOR data on muon bundles (dark points) with results of EAS measurements in AGASA [10], HiRes [11], Auger [12] and TA [13] experiments.

In figure 5, primary energy flux estimates derived from DECOR data are compared with results of recent measurements of PCR energy spectrum in the EeV energy range. Though a reasonable agreement in absolute normalization with AGASA data [10] is observed, it is seen from the figure that none of the examined interaction models allows us to match the data on muon bundles at large zenith angles with PCR spectrum measurements performed by means of the fluorescence method in HiRes [11], AUGER [12] and Telescope Array [13] experiments, even for the heavy (iron nuclei) mass composition.

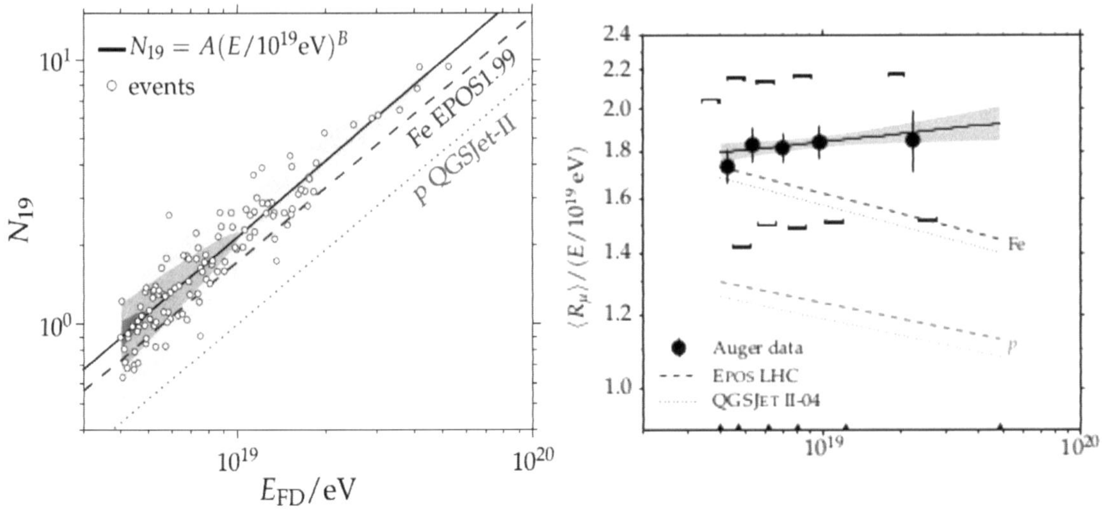

Figure 6. Excess of the number of muons in inclined EAS from The Pierre Auger Observatory data. a) Number of muons as measured by the fluorescence detectors, FD. [16] They concluded: *"We find that none of the current shower models, neither for proton nor for iron primaries, are able to predict as many muons as are observed b)* Average muon content $< R\mu >$ per shower energy E as a function of energy E. [17] they concluded *"We observe a muon deficit in simulations of (30-80)% at 10^{19} eV depending of the model".*

This contradiction becomes even more significant, if one takes into account that both HiRes and Auger data on X_{max} distribution favor a light (predominant proton) primary composition near 1 EeV [14,15]. If we assume proton primaries, the observed intensity of muon bundles will be about 2-3 times higher than the expectation. Recently, based on the analysis of the surface detector (SD) data on the muon content in inclined EAS simultaneously registered by means of the fluorescence detector (FD) the Auger collaboration came to the same conclusion that "none of the current shower models, neither for proton nor for iron primaries, are able to predict as many muons as are observed. [16,17]. This is shown clearly in figure 6.

The obtained results indicate with high probability the appearance of new processes of muon generation at very high energies that can explained by the appearance of new state of matter.

Acknowledgments
The author wishes to acknowledge prof. A. Petrukhin, spokesman of the experiment, and all his staff for his continuous encouragement to realize this experiment. I am also very grateful to Conacyt for the support trough the Mesoamerican Centre for Theoretically Physics.

References

[1] Aynutdinov, V. M. et al.; Astrophys. (1998) *Space Sci.* **258**, 105
[2] Amelchakov, M. B. et al.; (2001) *Proc 27th ICRC (Hamburg) vol 3 p 1267*
[3] Bogdanov, A. G. et. al.; (2010) *Phys Atom Nucl* **73** 1862
[4] Petrukhin, A. A. (for DECOR collaboration) *Nucl. Instru. Meth. Phys. Res* A **692** 228; and Saavedra O. (for DECOR collaboration) (2013) *Jour of Phys Conf Series* **409** 012009
[5] Heck, D. et. al.; (1998) *Forschunggszentrum Karlruhe Report* FZKA 6019
[6] Gaisser, T. K. and Stanev, T. *Particle Data Group* (2004) Phys Lett B **592** 228
[7] Yashin, I. I. et. al.; (2007) *Proc 30th ICRC (Merida Mexico)* vol 4 p 91
[8] Bertaina, M. et. al.; (2011) *Astrophys Space Scie Trans.* **7** 229
[9] Kokoulin, R. P. et. al.; (2009) *Nucl Phys* B *(Proc Suppl)* **196** 106
[10] Takeda, M. et. al.; (2003) *Astropart Phys* **19** 447
[11] Abbassi, R. U. et. al.; (2008) *Phys Rev Lett* **100** 101101
[12] Abraham, I. et. al.; (2010) *Phys Lett* B **685** 239
[13] Ikeda, D. et. al.; (2011) *Proc 32nd ICRC (Beijing, China)* vol 2 p 238
[14] Abbassi, R. et. al.; (2010) *Phys Rev Lett* **104** 161101
[15] Abraham, I. et. al.; (2010) *Phys Lett* **104** 091101
[16] Rodrigues, G. et. al.; (2013) *EPJ Web of Conf.* ***53*** *07003*
[17] A.Aab, et. al.; (2015) *Phys. Rev.* D ***91*** *032003*

IOP Conf. Series: Journal of Physics: Conf. Series **866** (2017) 012011 doi:10.1088/1742-6596/866/1/012011

Muon Lifetime Measurement in Chiapas and the Escaramujo project

Luis Rodolfo Pérez Sánchez[1] **and Federico Izraelevitch**[2]

[1]Faculty of Sciences in Physics and Mathematics and Mesoamerican Centre for Theoretical Physics of Autonomous University of Chiapas (FCFM-MCTP-UNACH)

[2]National University of San Martin, Argentina and Fermilab (Fermi National Accelerator Laboratory)

E-mail: rperez@mctp.mx

Abstract.
Escaramujo is a project with the goal of promote scientific development and integration regarding science for Latin America. It consists of a series of Laboratories and Workshops for High Energy Physics, astroparticle and instrumentation, given by Federico Izraelevitch. Escaramujo has been conduced at several institutions in Latin America. In this work, the moun mean lifetime measurements performed during the workshop held in Chiapas are presented. The results are compared with the corresponding value reported by the Particle Data Group (PDG).

1. Introduction

The activities performed during the workshop were based on a Cosmic Ray (CR) detector, designed specifically for the project. The students assembled, ran and controled the detector, they also performed data acquisition from several measurements. The detectors are designed specifically to provide a reasonably priced detection system on a scale which is suitable for classroom or personal use. After each course, a device remained in the academic institution to be used by teachers, students and researchers [1]. The principal goal, reported in this document, is to study the muonic component from an Extensive Air Shower (EAS), induced by CR.

The CR arrive to our planet and interact with the molecules in the atmosphere creating a shower of secondary particles. Those particles can be detectable at ground level by surface detectors or can be seen from their fluorescence traces on their way passing through the atmosphere, by means of Fluorescence telescopes and related techniques.

When primary CR, which are mainly protons (a relatively small fraction of them consist on heavier nuclei, such as Fe, He, Li, Si, N, photons, among others) produce secondary particles in the atmosphere, those particles are mainly pions ($\pi^{\pm}, \pi^{0}$) and kaons ($K^{\pm}, K^{0}$). Mouns ($\mu^{\pm}$) are produced by interaction of charged pions ($\pi^{\pm}$). The $\pi^{\pm}$ decay into a $\mu^{\pm}$ and its related neutrino (ν_{μ}) or anti–neutrino ($\bar{\nu_{\mu}}$), see equations 1 and 2.

The μ is a massive elementary particle that belongs to the second generation of leptons [2]. Its mass is 200 times the electron mass, and its mean lifetime is longer than other unstable particles. It does not interact with matter via the strong force but through the weak and electromagnetic forces and with gravity.

The μ^- decays into an electron (e^-), the μ^+ into positron (e^+) with its related neutrino (ν_l) and anti–neutrino ($\bar{\nu_l}$), see equations 3 and 4.

$$\pi^- \longrightarrow \mu^- + \bar{\nu_\mu} \quad (1)$$

$$\pi^+ \longrightarrow \mu^+ + \nu_\mu \quad (2)$$

$$\mu^- \longrightarrow e^- + \bar{\nu_e} + \nu_\mu \quad (3)$$

$$\mu^+ \longrightarrow e^+ + \nu_e + \bar{\nu_\mu} \quad (4)$$

1.1. Plastic Scintillator and Photodetector

The instruments used for this experiment are scintillation detectors, made by special materials that radiate when ionized and emit an small flash of light. This phenomenon occurs when the radiation interacts with matter by the excitation and ionization of a large number of atoms and molecules [2]. When they return to their ground state emits photons with energy range in the visible spectrum or near to it.

Escaramujo uses three plates of plastic scintillator *EJ-200* [1] which combines the two important properties: long optical attenuation length and fast timing, which makes it particularly useful for time-of-flight systems using scintillators greater than one meter long. Table 1 shows the properties of the plate EJ-200. The signal produced by the secondary particles in the scintillator are measured by Silicon Photomultipliers (*SiPM* [2]), one per plate.

The *EJ-200* and the *SiPM* are enclosed with EMI/Static shield (Type 1, Class 1) to exclude ambient light from getting into the SiPM, see figure 1.

Table 1. Properties of the plate EJ-200

PROPERTIES	EJ-200
Scintillation Efficiency (Photons/1 MeV e^-)	10000
Wavelength of Maximum Emission (nm)	425
Light Attenuation Length (cm)	380
Rise time (ns)	0.9
Decay time (ns)	2.1
Polymer Base	Polyvinyltoluene
Refractive index	1.58

The *SiPM* has a surface of 6 mm x 6 mm, type *MicroFC-60035-SMT-SensL* [3], this collects the light produced in the scintillator and data acquisition is performed with a high performance

[1] http://www.eljentechnology.com/index.php/products/plastic-scintillators/ej-200-e j-204-ej-208-ej-212
[2] Detector built from a photodiode matrix, working in avalanche mode, also known as Geiger Mode
[3] http://sensl.com/estore/microfc-60035-smt/

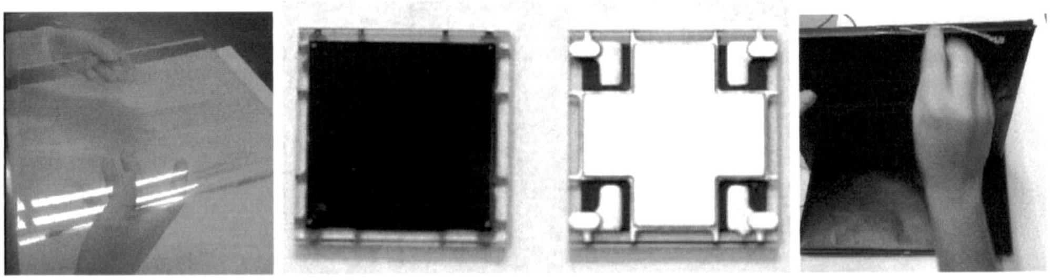

(a) (b) (c)

Figure 1. Components of Escaramujo: (a) EJ-200 size of 25 cm x 25 cm x 1 cm, before packaging, (b) Silicon photomulipliers of 6 mm x 6 mm, (c) Placing the *SiPM* in the scintillator after a careful packaging process

analog to digital converter (DAQ) (Quarknet board evaluation version 1.12). The SiPM are insensitive to the electromagnetic fields (EMF), it gives another option for the use of radiation detection and medical physics [3].

The DAQ card for the school-network CR detectors provides a low-cost alternative for its use in standard particle and nuclear physics fast pulse electronics modules, it produces trigger time and pulse edge time data for 2 to 4-fold coincidence levels. Individual detector stations (consisting of 4 scintillation counter modules, front-end electronics and a GPS receiver, produce an stream of data; ASCII text strings in a set of distinct formats for different functions). The DAQ has been designed with Leading and Trailing edge (relative) arrival times, measured with $\sim$ 1 ns precision [4].

The card includes a low-cost GPS (Ublox ANN-MS-2-005 GPS Antenna) receiver module, which allows time stamping for event triggers up to $\sim$ 50 ns absolute accuracy in UTC [5] between widely separated sites; the data are collected with a minicomputer Raspberry PI 2 model B.

2. Experimental Setup

Figure 2 shows the experimental setup for recording the pulse produced in the scintillator. Three plastic scintillator stacked gapless, on one side the SiPM adhere, BNC [4] cables were used to make the conections between the channels and the DAQ card, see Fig. 2 (a); the plates are not outdoor exposed, they are under concrete slabs from the building where the Escaramujo Detector (DE) is located. The concrete slabs have an approximate desity of $\sim$ 2.300 g/cm^3[6]. The EJ-200 plates, have a photomuliplier of silicon for detectors A, B and C. Figure 2 (b) shows an example of one of the most common detected events done with this configuration, namely a muon decay in Plate B.

The μ^- has a high probability of being captured by the positive charged nuclei of the detector. Here the events are coincidences between plates A and B or the trigger between A and B. The two neutrinos scape from the apparatus and, due to their low interaction probability, they will go directly through the Earth without stopping. The electrons emit light going through the scintillator [5].

[4] Bayonet Neill-Concelman

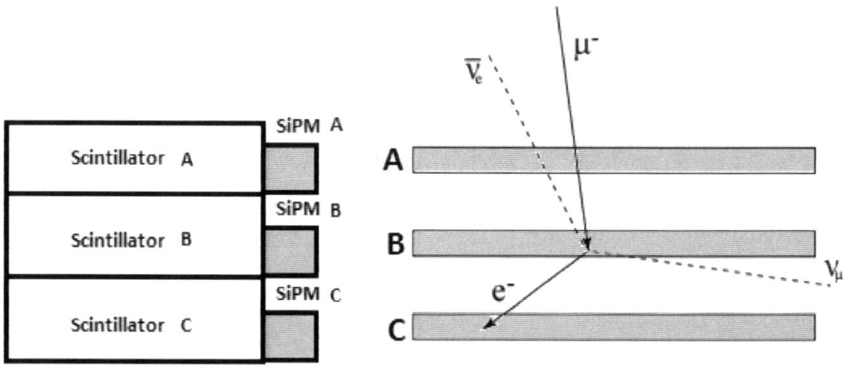

(a) (b)

Figure 2. (a)The experimental setup for recording the pulse and geometric configuration of the three scintillation detectors. (b) Plates A, B and C; μ decay on Plate B.

The experiment is an approach to get the μ mean lifetime. For this, after the detector assembly, data acquisition was done during 95 hours and 30 minutes. A Time Window (veto) to 400 ns, between the channel A and B was adjusted. Also a bias voltage of -30 mV and a threshold of 10 mV for the three channels has been used. The gate width is 40.960×10^3 ns, this represents the amount of time allowed between hits considered as coincident.

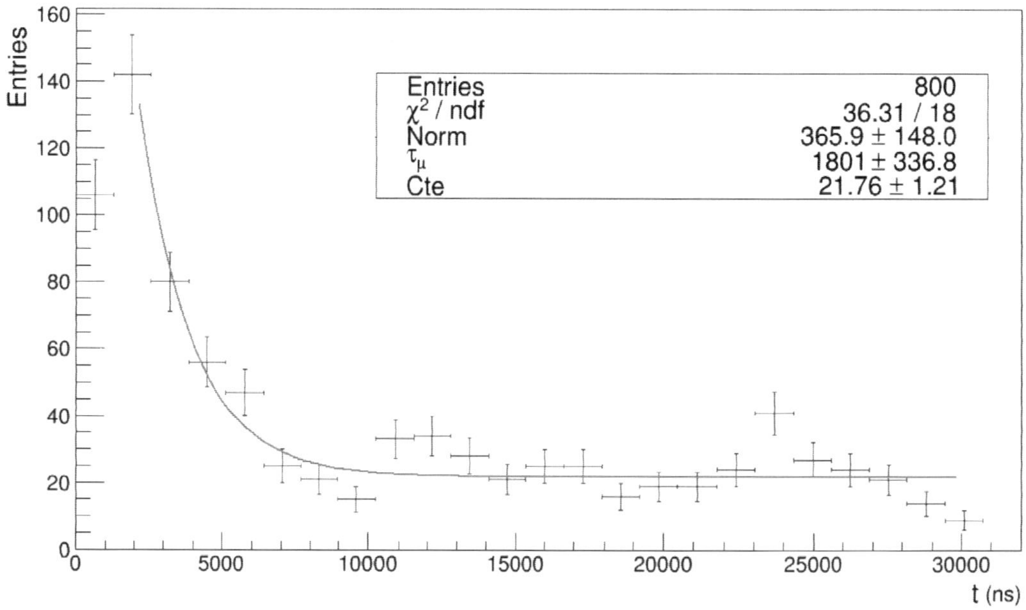

Figure 3. Muon decay. X-axis shows the time differences between the succeeding triggers, expressed in nanoseconds.

3. Results

To obtain the μ mean lifetime, its decay is used. It is supposed that for some time t an amount of $N(t)$ μ's is registered. If the probability for a μ to decay within some small time interval dt, is λdt, where λ is the dacay rate (a constant), the change dN is $dN = -N(t)\lambda dt$ or $dN/N(t) = -\lambda dt$. Integrating $\int_{N_0}^{N(t)} dN/N(t) = -\int_0^t \lambda dt$ it is obtained $N(t) = N_0 exp(-\lambda t)$, where $N(t)$ is the number of remaining ("survived") muons at time t and N_0 is the number of muons at $t = 0$. The moun mean lifetime (τ_μ) is inversely proportional to λ, $\tau_\mu = \frac{1}{\lambda}$. In the presented experiment a value of $\tau_\mu = 1801 \pm 366.8$ ns has been obtained, which is in a reasonable agreement with the value reported by the Particle Data Group[8]. To perform the corresponding muon decay histogram, a C++/Root [9] program was developed, the histogram is shown in Fig. 3.

4. Conclusions

The Escaramujo Project has been a success at UNACH. The measurement of the muon lifetime, performed in the City of Tuxtla Gutierrez, Chiapas, at an altitude of 522 m. a. s. l. is $\sim 1.80 \times 10^{-6}$ s, this result is in a good agreement with the value reported by the Particle Data Group of 2.197 ± 2.2^{-6} s. This corresponds to muons, $\mu^{\pm}$, which decay in Plate B due to the veto conditions between A and B. We achieved a correct assembly of the DE components, this means it can be used to calibrate the Cherenckov Detector (LAGO)[7], which is being developed in Chiapas.

5. Acknowledgment

Thanks to the Escaramujo Team in Chiapas, Mexico and to Dr. Arnulfo Zepeda (MCTP Director) for his unconditional support for all the scientific activities.

References

[1] *http://es.escaramujo.net/.*

[2] Glenn F. Knoll ***Radiation Detection and Measurement***. Professor of Nuclear Engineering and Radiological Sciences. University of Michigan.

[3] Erika Garutti, Peter Schleper. ***Study of the Silicon Photomultipliers and Their Applications in Positron Emission Tomography*** , zur Erlangung des Doktorgrades des Department Physik der Universität Hamburg.

[4] Sten Hansen Jordan, T. Kiper, Dan Claes, et. al, ***Low-Cost Data Acquisition Card for School-Network Cosmic Ray Detectors***, Fermi National Laboratory, Batavia IL, Univ. of Nebraska, Lincoln, NE. Univ. of Whashington, Seattle, WA.

[5] Jeff Rylander, T. Jordan, J. Paschke & Hans-Gerd Berns.***QuarkNet Cosmic Ray Muon Detector User's Manual Series "6000" DAQ***, Version 1.1, 1-53, January 2010.

[6] Donald E. Groom, Nikolai V. Mokhov & Sergei I. Striganov, ***Muon Stopping power and range Tables 10 Mev-100 Tev***, Atomic Data and Nuclear Data Tables **78**, 183356 (2001)

[7] M. Alarcón, F. Alcaraz, J. Barrera, A. Zepeda, et. al, ***Calibration and monitoring of water Cherenkov detectors with stopping and crossing muons***, Nuclear Instruments and Methods in Physics Research A 420 (1999) 3947.

[8] K. A. Olive *et al.* [Particle Data Group Collaboration], Chin. Phys. C **38**, 090001 (2014). doi:10.1088/1674-1137/38/9/090001

[9] *https://root.cern.ch/.*

IOP Conf. Series: Journal of Physics: Conf. Series **866** (2017) 012012 doi:10.1088/1742-6596/866/1/012012

Effects of the direct light in the surface detectors (SD) of the Pierre Auger Observatory and their change in time

Pedro Alfonso Valencia Esquipula[a], Karen Salomé Caballero Mora[a b].

[a]Facultad de Ciencias en Física y Matemáticas (FCFM). Universidad Autónoma de Chiapas (UNACH), Chiapas, Mexico.[b]Mesoamerican Centre for Theoretical Physics (MCTP).

E-mail: pa.valenciaesquipula@gmail.com

Abstract. Cosmic Rays (CR) are particles which come to the earth from Universe. Their origin and production mechanisms are still unknown. The Pierre Auger Observatory is located in Mendoza, Argentina. It is dedicated to the study of CR. When CR arrive to the earth's atmosphere they produce a shower of secondary particles called *air shower*. The surface detector (SD) of the Pierre Auger Observatory consists of tanks full of pure water, where CR produce *Cherenkov radiation*, when going through them. This light is detected by three photomultiplier tubes (PMT) located on the top of each tank. Depending of the angle of arrival direction of the primary CR, each PMT is able to register different signal than the other. The goal of this study is to look at these effects of direct light on the PMT's to explore if they change in time. The obtained results may give information about the physical status of the tanks in order to monitor the work of the SD, and to estimate possible systematic effects on the measurements. The current results of this study are shown.

1. Introduction

The Pierre Auger Observatory measures Cosmic Rays within an area of 3000 km^2. The design calls for 1660 water-Cherenkov detectors, arranged on a triangular grid, with the sides of the triangles measuring 1.5 km, overlooked by four optical stations, each containing six telescopes, designed to detect air-fluorescence light (Fig. 1). The water-tanks(stations) respond to the particle component (mainly muons, electrons and positrons) and the fluorescence cameras measure the emission from atmospheric nitrogen that is excited by the charged particles of the shower as they traverse the atmosphere. Both techniques, already used for many years to study extensive air showers, are brought together in a 'hybrid' detector to observe showers simultaneously with different techniques[1]. The array of water-tanks is known as the surface detector (SD) while the optical stations form the fluorescence detector, are known as fluorescence detector (FD).

IOP Conf. Series: Journal of Physics: Conf. Series **866** (2017) 012012 doi:10.1088/1742-6596/866/1/012012

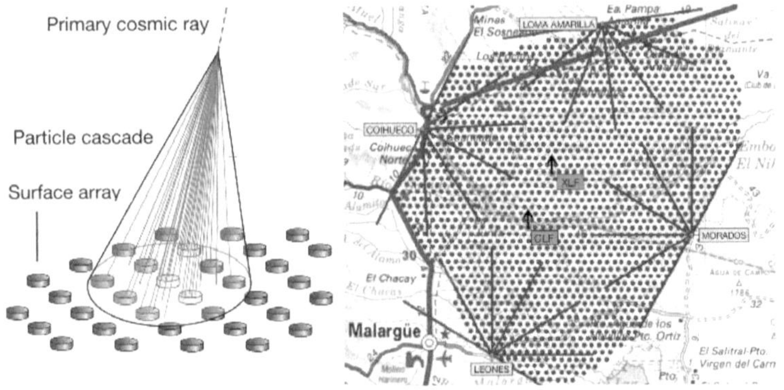

Figure 1. Representation of Air-Shower and Current status of the Pierre Auger Observatory.

2. Objective

The goal of this work is to observe the effects of direct light in the tanks which conform the SD, as well as their evolution as a function of time. The change in time could provide information on the physical state of each tank. Such information can be also used for studying possible systematic effects in the measurements made by the SD.

3. The Surface Detector

Each SD is a cylindrical water tank with $12,000$ L of pure water. It has a cross section of 10 m^2, a diameter of 3.6 m and a depth of 1.2 m (see Figures 2 and 3). The 1660 tanks of the SD are separated 1.5 km from each other. When an ultra-relativistic CR crosses the tank, it produces Cherenkov light, that is measured by three photomultipliers (PMTs) located on the top of each tank. Each PMT has $9 - in$ of diameter, these are Photonis XP1805. The PMTs are looking to the water, separated a distance of $1.2m$ from the center of the tank, having a $120°$ of difference in azimuthal angle from each other (see Figure 4).

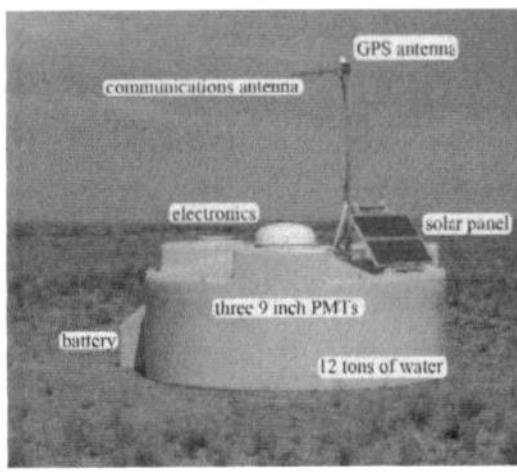

Figure 2. One Surface Detector station.

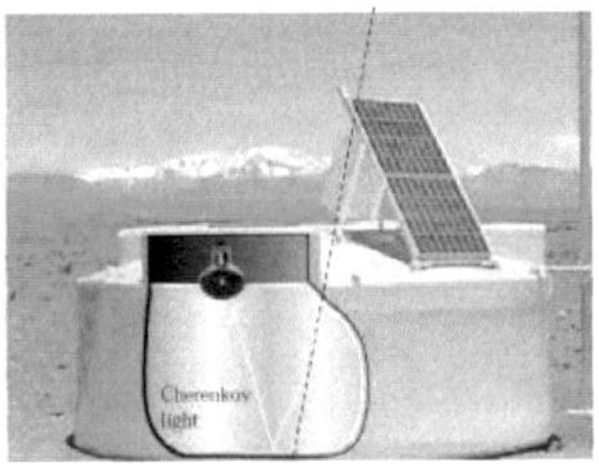

Figure 3. Particle crossing the tank.

4. Description of the effect of direct light

The direct light coming to each PMT is conditioned by the azimuthal angle of the arrival direction the particle which produced the light has. If the light arrives with the same azimuthal angle as the one a certain PMT has, then, the measurement made by that PMT will be higher than the one made by the other two. Such effect can be observed in a graph showing the signal as a function of azimuthal angle, which present a different maximum and phase for each PMT, as is shown in Graphs 1 to 4 [2].

IOP Conf. Series: Journal of Physics: Conf. Series **866** (2017) 012012 doi:10.1088/1742-6596/866/1/012012

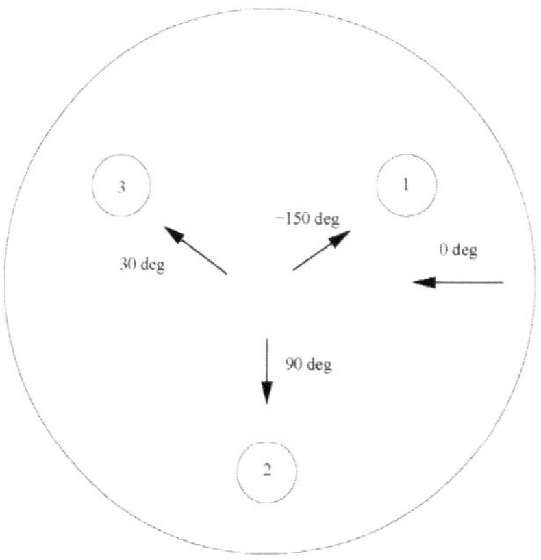

Figure 4. PMT positions on the top of the tank.

5. Description of the Analysis

A selection of tanks which have been working properly and more or less continuously over time, during the period from 2004 to 2013, has been made. Also quality cuts for events detected are requested to guarantee that the measurement corresponds to a well reconstructed event. The results shown in this document are obtained from the analysis of a tank named "Mariana" with number ID 338, which presented an event amount of 5673.

Each PMT data is plotted as a function of azimuth angle, and an approximation of the signal from each PMT is performed by adjusting a sine function of the azimuth angle with the form:

$$A\sin(\frac{\pi(x-k)}{180})+b) \tag{1}$$

With $k = 30$ for the PMT1. $k = 0$ for PMT2 and $k = 30$ for PMT3.

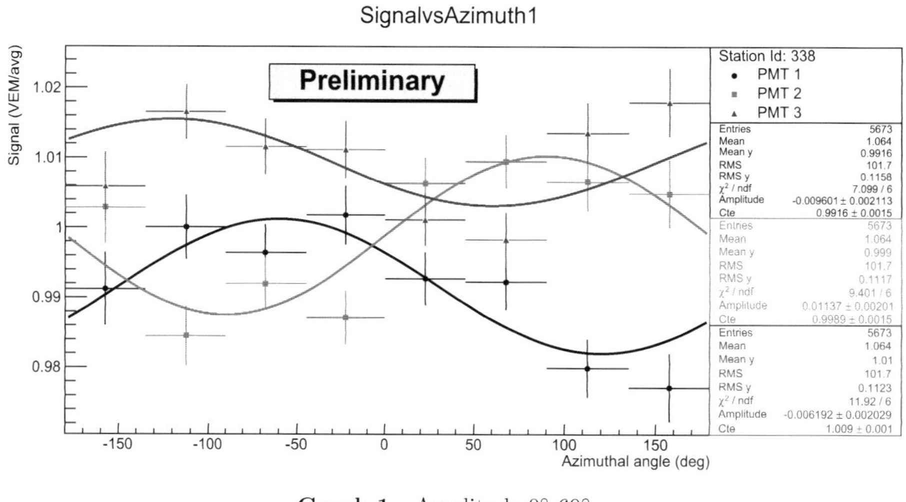

Graph 1. Amplitude 0°-60°.

Three zenith angle ranges, having the same solid angle, are considered: $0^\circ - 34^\circ, 34^\circ - 48^\circ$ and $48^\circ - 60^\circ$. The corresponding results, showing the azimuthal effect are shown in Graphs 2 to 4.

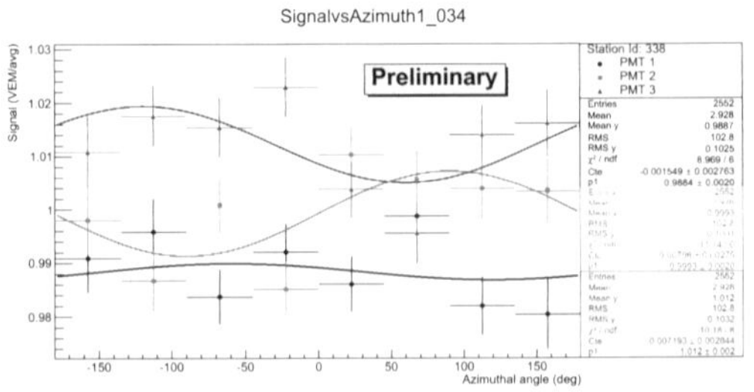

Graph 2. Amplitude 0°-34°.

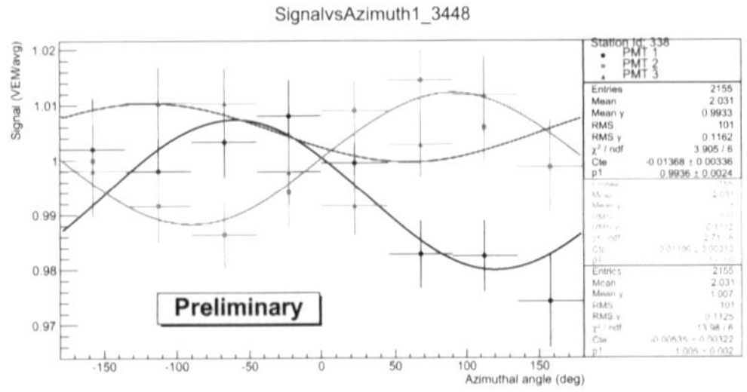

Graph 3. Amplitude 34°-48°.

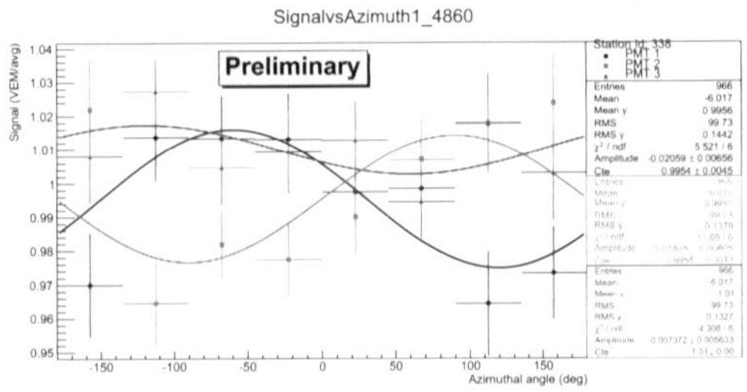

Graph 4. Amplitude 48°-60°.

The amplitudes of each PMT are plotted as a function of zenith angle. The corresponding line is fitted with a function of the form:

$$y = mx + b \tag{2}$$

Results for PMT 1, PMT 2 and PMT 3 are shown in Graphs 5 -7.

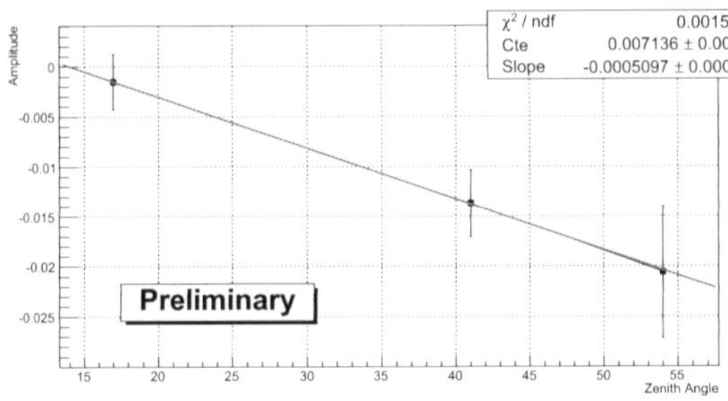

Graph 5. Amplitude adjustment for the PMT1.

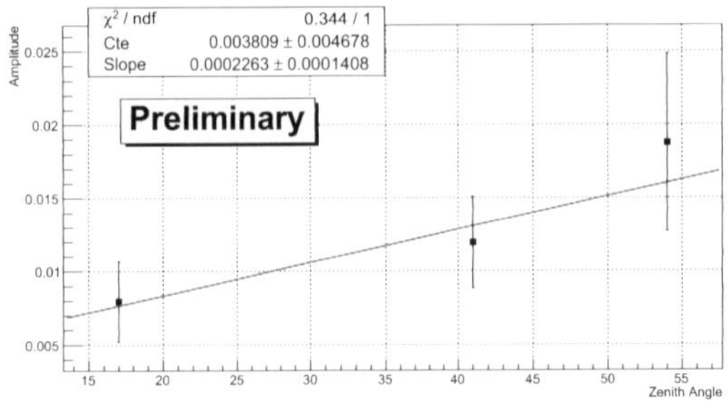

Graph 6. Amplitude adjustment for the PMT2.

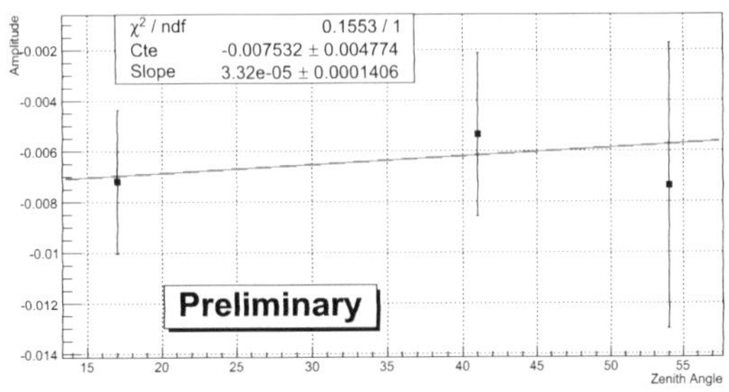

Graph 7. Amplitude adjustement for the PMT3.

In order to obtain only one characteristic line for each detector, the amplitudes for each PMT at a certain zenith angle are averaged and plotted as shown in Graph 8. Then a new linear fit is performed resulting in the corresponding characteristic line.

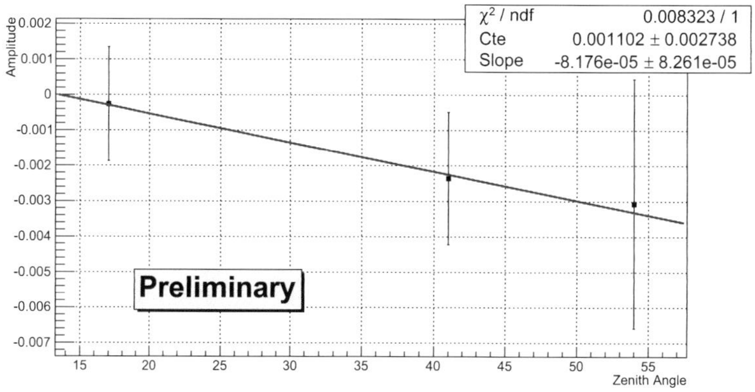

Graph 8. Characteristic line of the station.

Once the characteristic line of the station was obtained for the whole period of time, its change in time is to be explored. The same study has to be done for three different periods in order to compare the corresponding characteristic lines. This is work in progress.

6. Conclusion

From graphs 1-4, it can be seen that the effect of direct light is effectively present for each PMT of of one station in the SD. A characteristic line, for the station studied was found. The next step is to explore signals of the same tank for different time periods in order to study a possible change of the characteristic line as a function of time. If some change is found, its impact in the measured signal has to be studied. The same kind of study must also be performed for other tanks.

7. Aknowledgements

- To CONACyT for the support given by the project CB 243290.
- To L'ORÉAL Mexico, CONACyT, UNESCO, CONALMEX and AMC fort the support granted through the Fellowship for Women in Science 2014 to Karen Salomé Caballero Mora.

8. References

1 The Pierre Auger Collaboration, arXiv:1604.03637v1, astroph-IM, The Pierre Auger Observatory Upgrade Preliminary Design Report (2016).

2 K.S. Caballero-Mora, L. Nellen and J.F. Valdes-Galicia. *Azimuthal signal variations in the engineering array of the Pierre Auger Observatory.* Revista mexicana de Física Ag 2008 54 (4) 306-313

IOP Conf. Series: Journal of Physics: Conf. Series **866** (2017) 012013 doi:10.1088/1742-6596/866/1/012013

Cosmic-ray physics at CERN

M. Rodríguez Cahuantzi

Faculty of Physics and Mathematics, Autonomous University of Puebla
Av. San Claudio y 18 Sur, Edif. EMA3-231, Ciudad Universitaria 72570, Puebla, México

E-mail: mrodriguez@fcfm.buap.mx / mario.rodriguez@correo.buap.mx

Abstract.
Accelerator experiments located underground are suitable for the study of atmospheric muons. The use of high-energy collider detectors for cosmic-ray physics was pioneered during the era of the Large Electron-Positron (LEP) collider at CERN by ALEPH, DELPHI and L3 collaborations. A development of these programs is possible at the Large Hadron Collider (LHC), where experiments like ALICE and CMS will operate for many years, with the possibility of recording a large amount of cosmic-ray data. In this proceedings, a review of the results obtained by LEP and LHC experiments is presented. This material was discussed along two sessions during the VI School on Cosmic-ray Physics and Astrophysics held at the Mesoamerican Center for Theoretical Physics (MCTP) located in Tuxtla Gutierrez, Chiapas, Mexico.

1. Introduction to accelerators and particle detectors

Since the invention of particle accelerators in the 1930s decade, they have been used for different purposes: basic research, applied sciences and industry. The energy of a beam of particles is increased by accelerators using electromagnetic fields to boost them to nearly the speed of light. At the same time, the total momentum of the particles is increased (wavelength decreases) and thus it is possible to inquire into the inside the atoms. The acquired energy by the particles within accelerators allows the production of massive particles which are studied by different experimental techniques involving the development of sophisticated particle detectors.

During the twentieth century, several accelerators complexes have been developed throughout the world to be used as colliders to carry out studies in particle physics (see table 1).

Table 1. List of some accelerator complexes

Name	Location	Main studies
SLAC	California-USA	discovery of charm quark and τ lepton
Fermilab	Illinois-USA	discovery of bottom and top quarks and τ neutrino
CERN	Swiss-France border	discovery of Z, W particles and Englert-Brout-Higgs Boson
BNL	NY-USA	discovery of charm quark simultaneously with SLAC
CESR	NY-USA	detailed studies of bottom quark
DESY	Hamburg-Germany	discovery of gluons
KEK	Tsukuba-Japan	B-mesons studies
IHEP	China	τ lepton and charm quark studies

Particle accelerators can be constructed as circular or linear arrangements to be able to provide two types of collisions:

- *fixed target*: accelerated charged particles collide with a specific target made by a solid, liquid or gas. The particles are produced in the forward direction. The detectors are cone shaped and placed downstream.
- *colliding beams*: two accelerated beams of charged particles collide by making them crossing to each other. Because the particles are produced in all directions (higher mass particles are easily produced in colliding beam arrays), most of the detectors are cylindrical or spherical.

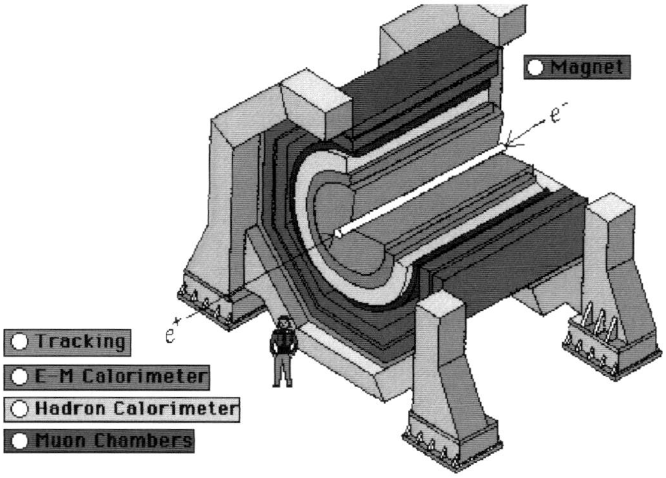

Figure 1. Schematic view of a particle detector arrangement. Figure taken from [1]

A view of a typical arrangement of particle detectors used in circular colliding beams is shown in figure 1. This type of detectors are usually constructed in four slices accordingly to the different particle species to be identified:

- *Tracking:* used to reconstruct the trajectories (tracks) and vertexes of the charged particles.
- *Electromagnetic calorimeter:* used to measure the energy of e^+, e^- pairs and photons
- *Hadron calorimeter:* used to measure the total energy of the produced hadrons.
- *Muon chambers:* used to detect muons. Neutrinos can be reconstructed as *missing energy.*
- *Magnet:* used to deflect the trajectories of produced charged particles. The radius of curvature is helpful to reconstruct the momentum and charge of the particles.

Electrons and protons are detected both in the tracking chamber and electromagnetic calorimeters. Neutral particles (neutrons and photons) can not be reconstructed by the tracking chambers. Photons are detected by the electromagnetic calorimeter and neutrons are evidenced by the deposited energy in the hadron calorimeter [1].

2. Contribution of accelerators to cosmic-ray physics

Cosmic-rays (CR) are conformed by a beam of particles generated outside the Earth's atmosphere. The composition of primary cosmic-rays is 90% protons, 9% **He** nuclei and 1% heavier nuclei. The energy range of primary cosmic-rays that reach our planet starts from a few GeV per particle up to 10^{20} eV. At low energies the composition of the primary cosmic rays tends to be formed by protons with a few percentage of **He** nuclei, while at higher energies the nuclear contribution increases.

We can see in table 2 a clear contribution from the CR community to the discovery of elementary particles since early 1930's decade.

Table 2. Discovery of elementary particles

Particle	Year	Discovered	Technique
e^-	1897	Thomson	Discharge in gases
p	1919	Rutherford	Radioactivity
n	1932	Chadwik	Radioactivity
e^+	1933	Anderson	Cosmic-rays
$\mu^{+/-}$	1937	Neddermeyer, Anderson	Cosmic-rays
$\pi^{+/-}$	1947	Powell, Occhialini, Lattes	Cosmic-rays
$K^{+/-}$	1947	Rochester, Butler	Cosmic-rays
π^0	1949	Bjorklund	Accelerator
K^0	1951	Armenteros	Cosmic-rays
Λ^0	1950	Hopper	Cosmic-rays
anything else	1955 $\rightarrow$ today	various groups	Accelerators

In recent decades, the extensive air showers (EAS) created by the collisions between the primary cosmic-rays and the atmosphere nuclei components have been studied by several experimental arrays around the globe. The precise measurements of the energy spectrum of the primary cosmic-rays for energies between $10^{15} eV$ and 10^{17} eV bring us information related with their origin and acceleration mechanisms. The composition of the primary cosmic-rays for energies below 10^{14} eV can be measured directly by using satellites and balloons. For higher energies, large ground and underground experimental setups are used.

The interpretation of the data collected by dedicated cosmic-rays experiments depends on the extrapolation of theoretical models beyond the energy range where were they adjusted or they were tested for particle production. While some of the models converge in a common interpretation, these may lack some predictability at higher energies. Until now there are not experimental data from accelerators regarding the particle production at small angles in the forward region at energies around the region of the cosmic ray spectrum known as the *knee* (10^{15} eV). Furthermore, new physical phenomena in a very forward region at high energies in hadronic interactions, such as coherent photo-production of vector mesons or heavy flavor production, can significantly influence the measurements done with ground and underground experimental arrays. It is therefore necessary to carry out as much as possible measurements to achieve a good understanding of the forward production of particles and thus of the primary cosmic-rays flux energy. It is of particular interest the study of momentum of atmospheric muons as well as high muon multiplicity events collected with underground apparatus.

IOP Conf. Series: Journal of Physics: Conf. Series **866** (2017) 012013 doi:10.1088/1742-6596/866/1/012013

In a typical cosmic-ray experiment, the detection of atmospheric muons is usually done using large-area arrays on the surface of the Earth or with detectors deep underground. The main purpose of such experiments is to study the mass composition and energy spectrum of primary cosmic rays in an energy range above 10^{14} eV, which is not available through direct measurements using satellites or balloons. The big advantages of these apparatuses is the large size and, in the case of surface experiments, the possibilities for measuring different particles, such as electrons, muons and hadrons, created in extensive air showers. Because the detectors involved in collider experiments are tiny compared with the large-area arrays, the approach and the studies have to be different so that the remarkable performances of the detectors can be exploited [2].

3. LEP results

The use of high-energy physics detectors for cosmic-ray physics was pioneered by ALEPH (Cosmo-ALEPH) [3], DELPHI [4] and L3 (L3+C) [5] experiments during the Large Electron-Positron (LEP) collider era at CERN. In table 3 a summary of the topics in CR physics covered by each collaboration is shown.

Table 3. LEP experiments which contributed to Cosmic-ray physics. Table taken from [6]

Detector	Depth (m)	Topics
Cosmo-ALEPH	130	muon energy spectrum, μ^+/μ^- charge ratio, muon bundles, sources
DELPHI	100	muon bundles, source map
L3+C	50	angular dependence, μ^+/μ^- charge ratio, absolute muon spectrum

The main contribution of LEP experiments to the cosmic-ray physics was the measurement of the atmospheric muon momentum spectra and the detection of multi-muon bundles by the central barrel detectors. These results helped in the understanding of hadronic interaction models as well as the constrains of atmospheric neutrino fluxes. By measuring the atmospheric muon multiplicity distribution (MMD), LEP experiments concluded that the bulk of the data can be successfully described using standard hadronic production mechanisms, but the events with more than 100 reconstructed atmospheric muons occur with a frequency which is almost one order of magnitude above the simulation [7], even when assuming a combination of an extreme heavy composition with the higher measured flux values [8].

A nice and detailed discussion on LEP results on this topic can be found in [6].

4. LHC results

The Large Hadron Collider (LHC) is the largest particle accelerator ever built. It is placed in the same tunnel as LEP. The conditions at LHC allow the detailed study of $p-p$, $Pb-Pb$ and $p-Pb$ collisions:

- Large Quark Gluon Plasma temperature, volume, energy density and lifetime,
- Large cross section for hard probes: high p_T, jets, heavy quarks,
- Small net-baryon density at mid-rapidity corresponding to the conditions of early Universe,
- First principles methods more applicable (pQCD, Lattice Gauge Theory),
- New generation of particle detectors: ALICE, ATLAS, CMS and LHCb

A development of CR Physics programs is possible at the LHC, where experiments operate for many years, with the possibility of recording a large amount of cosmic-ray data. In this context, the ALICE [9] and CMS [10] collaborations began separated programs to collect cosmic-ray data during pauses in LHC operations (see table 4).

Table 4. LHC experiments with Cosmic-ray physics programs.

Detector	Depth (m)	Topics
ALICE	50	multi-muon bundles, μ^+/μ^- charge ratio
CMS	89	μ^+/μ^- charge ratio

4.1. CMS contribution to CR Physics
The Compact Muon Solenoid (CMS) is a particle detector designed to study high energy $p-p$ collisions at LHC. It is composed by four slices: silicon tracker, calorimeters (electromagnetic and hadronic), superconducting solenoid and iron return yoke interspersed with muon chambers (see figure 2). The CMS apparatus is located 89 m. underground with 50 m. of moraines and 20 m. of molasse rock. The solenoid magnet of CMS provides a field of 4 teslas which allows the reconstruction of the momentum of the atmospheric muons that reach the central barrel of CMS spectrometer [10].

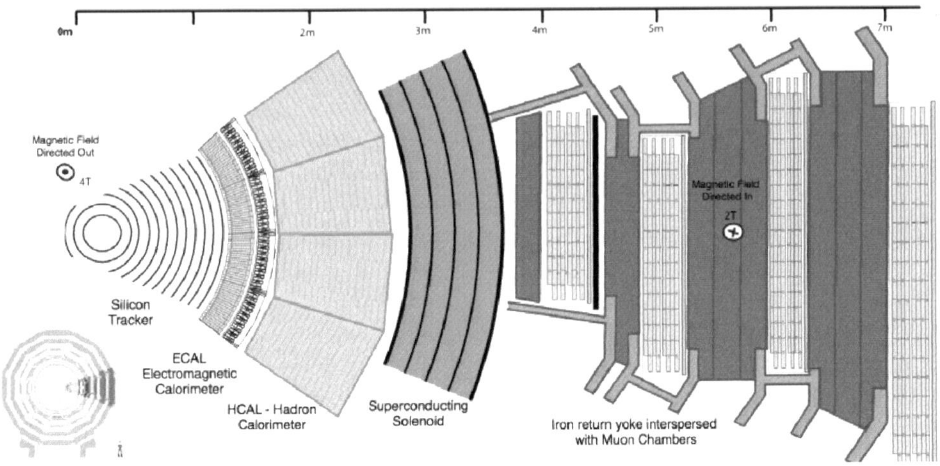

Figure 2. Schematic view of CMS detector. Figure taken from [11]

The CMS collaboration used the cosmic-ray data collected during the commissioning of the detector to carry out studies related with the measurement of μ^+/μ^- cosmic charge ratio. The muon charge ratio reflects the excess of π^+ over π^- and K^+ over K^- in the forward fragmentation region of proton initiated interactions together with the fact that there are more protons than neutrons in the primary spectrum. The increase with energy of this quantity reflects the increasing importance of kaons in the TeV range and indicates a significant contribution of associated production by cosmic-ray protons. The same process is even more important for atmospheric neutrinos at high energy [12].

During the cosmic data taking in 2006 and 2008, CMS recorded around 295 million of events with a magnetic field ranging from 3.67 Teslas to 4 Teslas. The experimental data were treated in three different ways:

- *stand-alone muon track* (MTCC): reconstruction based only on hits from the muon chambers,
- *tracker track* (STA): reconstruction based only on hits from the silicon tracker,
- *global-muon track* (GLB): reconstruction based on hits from muon chambers and silicon trackers.

In figure 3 the results of each individual analysis are shown. The sources of systematic uncertainties are: trigger efficiency, charge misalignment and material description. In the region where the results overlap, CMS made a combination of them. For a momentum less than 100 GeV/c, the μ^+/μ^- cosmic charge ratio is 1.2766 ± 0.0032 (stat.) ± 0.0032 (syst.) which is in good agreement with the measurements made by previous experiments [13, 14, 15, 16].

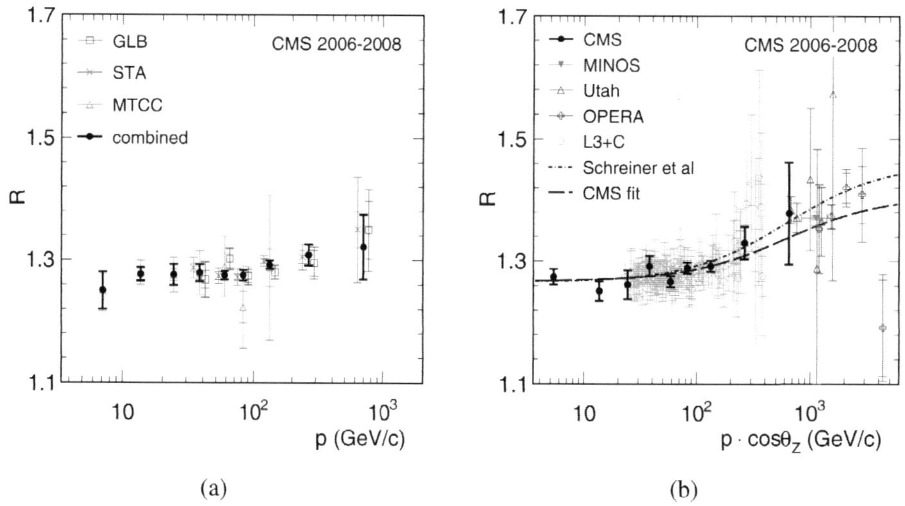

Figure 3. (a) The three CMS results, and their combination, as a function of the muon momentum. Data points are placed at the bin average, with the points from the standalone and global-muon analyses offset horizontally by ±10% to allow comparisons. (b) The CMS result, as a function of the vertical component of the muon momentum, together with some previous measurements and a fit of the pionkaon model to the CMS data [12]

CMS collaboration measured the flux ratio of positive to negative charge cosmic-ray muons as a function of the momentum and its vertical component. The result reported by CMS is in excellent agreement with previous measurements made by underground experiments. This is the most precise measurement of the charge ratio in the momentum region below 0.5 TeV/c. It is also the first measurement using muons with the complete CMS detectors. Details on this can be found in [12].

4.2. ALICE contribution to CR Physics

ALICE is one of four large experiments at the CERN Large Hadron Collider. Located 52 meters underground with 28 meters of overburden rock, it has also been used to detect atmospheric muons produced by cosmic-ray interactions in the upper atmosphere. In this section, the muon multiplicity distribution of these cosmic-ray events and their comparison with Monte Carlo simulation is discussed. The analysis made by ALICE exploits the large size and excellent tracking capability of its Time Projection Chamber (TPC). They gave a special emphasis to the study of high multiplicity events containing more than 100 reconstructed muons and corresponding to a muon areal density larger than 5.9 m^{-2}. The ALICE measurement on the rate of these events shows that they stem from primary cosmic-rays with energies above 10^{16} eV. The frequency of these events can be successfully described by assuming a heavy mass composition of primary cosmic-rays in this energy range and using the most recent hadronic interaction models to simulate the development of the resulting air showers.

ALICE is located at Point 2 of the LHC accelerator tunnel, in a cavern placed 52 m underground and with 28 m overburden of rock, at 450 m above the sea level. The rock absorbs all of the electromagnetic and hadronic components of an EAS, while near vertical muons with a surface energy greater than 16 GeV can reach the detectors. ALICE is a typical collider experiment, with a solenoid magnet housing the central barrel detector, and a forward muon arm, consisting of absorbers, a large dipole magnet, and fourteen planes of tracking and triggering chambers located outside the ALICE magnet. A complete description of the apparatus is given in [9].

The ALICE TPC was used to reconstruct the trajectory of cosmic-ray muons passing through the active volume of the detector. For the purpose of detecting cosmic-ray muons, the effective area of the detector considering its horizontal cylindrical geometry is approximately 17 m^2.

Specific triggers have been implemented to detect atmospheric muons crossing the central barrel of the ALICE apparatus (see Fig.4). To trigger and measure cosmic-ray events, three detectors were used: ACORDE (Alice COsmic Ray DEtector), TOF (Time of Flight) and SPD (Silicon Pixel Detector).

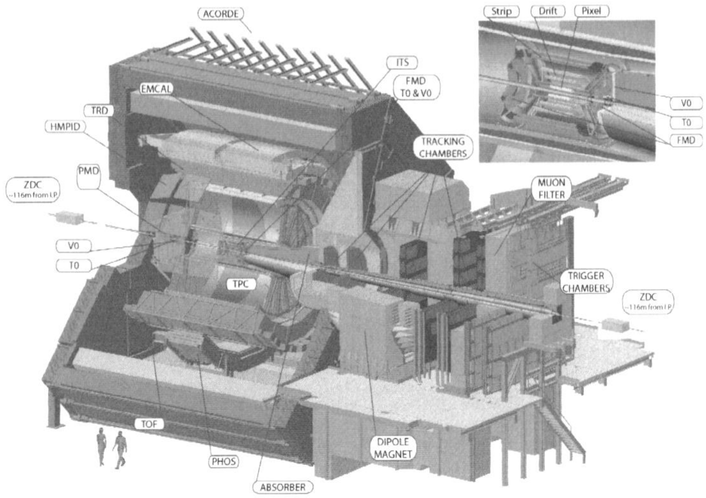

Figure 4. The ALICE detector at P2-LHC. ACORDE, TOF, TPC and SPD are used to trigger and measure cosmic-ray events.

ACORDE consists of an array of 60 scintillator modules located on the three top octants of the ALICE magnet. Each module is composed of two superimposed plastic scintillator paddles with an effective detection area of 0.37 m^2 (see Fig. 5). The trigger is given by the coincidence of the signals in n different modules (n-fold coincidence) in a 100 ns time window. A typical

IOP Conf. Series: Journal of Physics: Conf. Series **866** (2017) 012013 doi:10.1088/1742-6596/866/1/012013

configuration of ACORDE is $n = 4$.

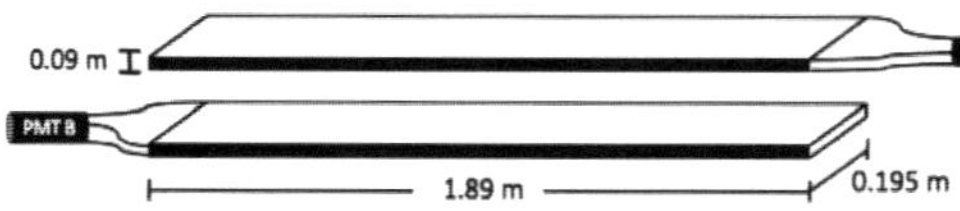

Figure 5. Schematic view of one ACORDE module. The light produced in the scintillator plastic is driven by two optical guides to one PMT per paddle. The information of the pulse is sent to the FEE (front end electronics) to generate the ACORDE cosmic-ray trigger.

The SPD is part of the Inner Tracking System located inside the inner field cage of the TPC. It is composed of two layers of silicon pixel modules located at a distance of 39 mm and 76 mm from the LHC beam axis, respectively. The layers have an active length of 28.3 cm, centered upon the nominal interaction point of the LHC beams. The SPD was incorporated into the trigger by requiring a coincidence between signals in the top and bottom halves of the outermost layer.

TOF detector is a cylindrical MRPC (Multi-gap Resistive Plate Chamber) array, completely surrounding the TPC. The trigger requires a signal in a read-out channel (a pad) in the upper part of the TOF and another in a pad in the opposite lower part forming a back-to-back coincidence with respect to the central axis of the detector.

A logical OR among the trigger signals of ACORDE, TOF and SPD was configured to generate the cosmic-ray trigger of ALICE. Most events were classified as either single muon events or multi-muon events, with a small percentage of "interaction" events where very energetic muons have interacted with the iron yoke of the magnet producing a shower of particles that pass through the TPC.

The atmospheric muons that cross the ALICE central barrel are reconstructed as two tracks (see Fig. 6). In order to quantify the muon multiplicity per event, a dedicated algorithm was developed by ALICE to match both reconstructed tracks and to obtain the full length of the muon track crossing the TPC (*up* and *down* tracks). However, the tracking algorithm has not been optimized for very inclined (quasi horizontal) tracks. Therefore, to avoid reconstruction inaccuracies associated with the most inclined showers, ALICE restricted the zenith angle of all events to the range $0^\circ < \theta < 50^\circ$.

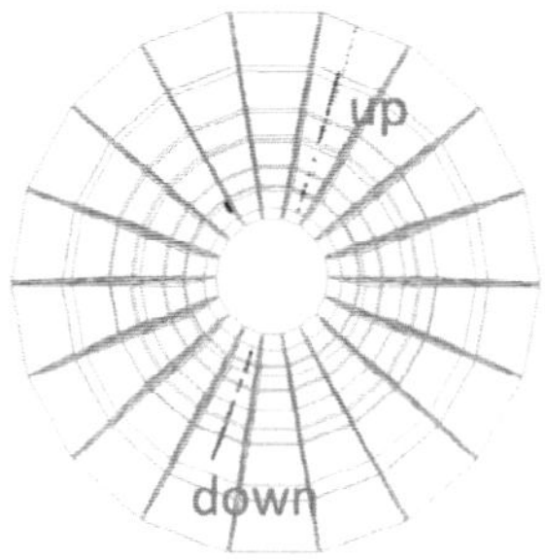

Figure 6. Track reconstruction of one single atmospheric muon with the TPC of ALICE.

Each TPC track can be reconstructed with up to 159 clusters (space points). In this analysis, tracks with at least 50 clusters in the TPC and a momentum larger than 0.5 GeV/c were required.

Since the atmospheric muons coming from the same EAS event arrive almost parallel at ground level, all the analyzed tracks per event should be parallel between them.

Finally, each *up-track* is matched to the closest *down-track* when the distance in the transverse plane (XZ) between them is less than 3 cm. A muon reconstructed with two tracks is called "matched muon". If the matching condition is not fulfilled the track is anyway accepted as a muon and called "single-track muon". Most of the single-track muons originate from particles crossing the TPC close to its edge, where part of the muon trajectory may fall outside the detector.

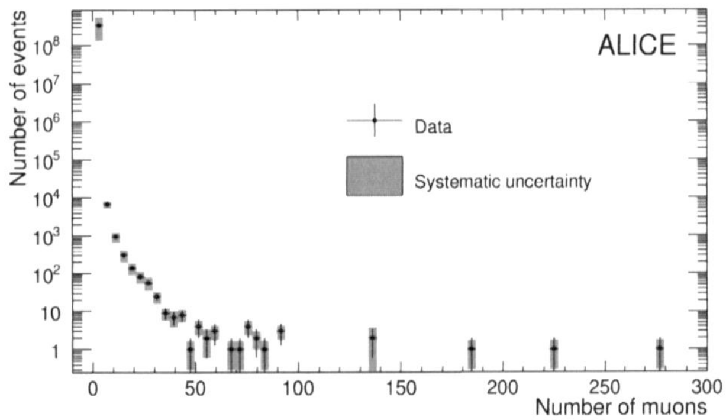

Figure 7. Atmospheric-muon multiplicity distribution taken from [17]

The main topic related to cosmic-ray physics investigated by ALICE is the study of the muon-multiplicity distribution (MMD) and in particular of events with high density of muons. The MMD measured by ALICE is shown in Fig. 7. There are 5 events with a number of muons greater than 100. As expected, a smooth distribution can be observed up to a muon multiplicity around 70. The MMD reconstructed by ALICE is very similar to that obtained by the LEP experiments [4, 5], which could not be explained by Monte Carlo models.

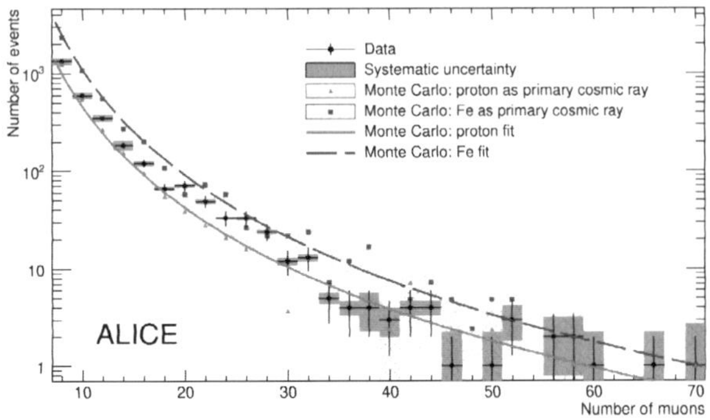

Figure 8. Atmospheric-muon multiplicity distribution of the data compared with the fit obtained with CORSIKA 6990 for proton (red line) and Fe (blue line). Figure taken from [17]

The data points shown in figure 8 are, as expected, in between the pure proton composition (light elements) and pure Fe (heavy elements). The lower multiplicities (lower primary

IOP Conf. Series: Journal of Physics: Conf. Series **866** (2017) 012013 doi:10.1088/1742-6596/866/1/012013

Table 5. Comparison of the HMM event rate obtained with the full simulation and from measurement. Table taken from [17]

HMM events	CORSIKA 6990 QGSJET II-03		CORSIKA 7350 QGSJET II-04		Data
	proton	iron	proton	iron	
Period [days per event]	15.5	8.6	11.6	6.0	6.2
Rate [$\times 10^{-6}$ Hz]	0.8	1.3	1.0	1.9	1.9
Uncertainty (%) (syst + stat)	13	16	8	20	49

energies) are closer to pure proton. At lower multiplicities, corresponding to lower primary energies, ALICE reported that the data approach the proton curve, which represents a light ion composition of the primary cosmic ray flux, while higher multiplicity data lie closer to the iron curve, representing a heavier composition. The limited statistics for a number of muons larger than 30 does not allow for a precise, quantitative study of the composition, but the distribution below this multiplicity suggests that the average mass of the primary cosmic ray flux increases with increasing energy [17].

Events with more than 100 atmospheric muons (HMM) reconstructed by the ALICE-TPC are due to primary cosmic-rays with an energy larger than 10^{16} eV [18]. To estimate the rate of the high muon-multiplicity events with the Monte Carlo models, ALICE simulated one year of effective data taking with Corsika 6990 (QGSJET II-03) and Corsika 7350 (QGSJET II-04, calibrated with TOTEM results on $\sigma_{pp}^{tot/el}$ [19]):

- energy of the primary cosmic-ray: $10^{16} < E < 10^{18}$ eV,
- zenith angle range $0^0 - 50^0$

The fluctuations of these rare events are relatively small in the Monte Carlo (which corresponds to 1 year of data) but are quite large in the real data (31 days of data taking). Within the large uncertainties quoted in table 5, there is agreement between the HMM rate observed in data and in the Monte Carlo calculation which makes use of CORSIKA supposing a heavy composition at these energies ($E > 10^{16}$ eV) [18].

5. Conclusions

LEP experiments provided important results in the field of cosmic-rays physics. The muon spectra obtained by Cosmo-ALEPH and L3+C helped to better constrain atmospheric neutrino fluxes and also to improve the knowledge of the ν_μ neutrino induced background in neutrino astronomy telescope. Measurements on multi-muon events were used to test hadronic interactions models at high energies: at low multiplicities a light component of the primary cosmic ray is favored while at higher multiplicities the interaction models failed to describe the data [6].

LHC experiments are continuously taking cosmic-ray data since 2006. The CMS experiment measured the μ^+/μ^- cosmic charge ratio as a function of the muon momentum and its vertical component. The result is in agreement with previous measurements by underground experiments. This is the most precise measurement of this quantity in the momentum region below 0.5 TeV/c [12].

On the other hand, the ALICE experiment collected about 31 days of dedicated cosmic-ray data. The analysis of the collected data on the measurement of MMD and its comparison with the Monte Carlo simulations suggest a mixed composition with an increasing average mass of the primary cosmic ray at higher energies. The obtained MMD is in agreement with most experiments working in the energy range around the knee. Using CORSIKA 6990/7350 (QGSJET II-03/04) as hadronic interaction model, ALICE was able to simulate these events and to reproduce,within relatively large uncertainties, their rate. It seems that most of the HMM are due to iron or heavy nuclei with an energy greater than 10^{16} eV and a shower core located near ALICE [18]. This is the first time that the rate of HMM events, observed at the relatively shallow depth of ALICE, has been satisfactorily reproduced using a conventional hadronic model for the description of extensive air showers; an observation that places significant constraints on alternative, more exotic, production mechanisms.

5.1. Acknowledgements
This work was supported by CONACyT México Basic Science projects under grant number 24408 (CB-2015-2018) and SEP-PRODEP project under grant number PTC-431, DSA/103.5/15/7449. The author also thanks to the organizers of the VI School on Cosmic-ray Physics and Astrophysics held in the Mesoamerican Center for Theoretical Physics (MCTP) for their kind invitation.

References
[1] *The Particle Adventure web page*
[2] Alessandro B and Rodriguez M 2012 *CERN Cour.* **52N6** 26–27
[3] Grupen C, Kurt M, Mailov A, Mller A, Putzer A, Rensch B, Sander H, Schmeling S, Schmelling M, Wachsmuth H, Ziegler T and Zuber K 2003 *Nuclear Instruments and Methods in Physics Research Section A: Accelerators, Spectrometers, Detectors and Associated Equipment* **510** 190 – 193 ISSN 0168-9002 proceedings of the 2nd International Symposium on Applications of Particle Detectors in Medicine, Biology and Astrophysics URL http://www.sciencedirect.com/science/article/pii/S0168900203016978
[4] Ridky J and Travnicek P 2005 *Nuclear Physics B - Proceedings Supplements* **138** 295 – 298 ISSN 0920-5632 URL http://www.sciencedirect.com/science/article/pii/S0920563204006310
[5] Stickland D P 1991 *Philosophical Transactions: Physical Sciences and Engineering* **336** 223–236 ISSN 09628428 URL http://www.jstor.org/stable/53785
[6] Travnicek P 2006 *Czechoslovak Journal of Physics* **56** A185–A200 ISSN 1572-9486 URL http://dx.doi.org/10.1007/s10582-006-0154-y
[7] Avati V, Dick L, Eggert K, Strm J, Wachsmuth H, Schmeling S, Ziegler T, Brhl A and Grupen C 2003 *Astroparticle Physics* **19** 513 – 523 ISSN 0927-6505 URL http://www.sciencedirect.com/science/article/pii/S0927650502002475
[8] DELPHI 2007 *Astroparticle Physics* **28** 273 – 286 ISSN 0927-6505 URL http://www.sciencedirect.com/science/article/pii/S0927650507000916
[9] ALICE 2008 *Journal of Instrumentation* **3** S08002 URL http://stacks.iop.org/1748-0221/3/i=08/a=S08002
[10] CMS 2008 *Journal of Instrumentation* **3** S08004 URL http://stacks.iop.org/1748-0221/3/i=08/a=S08004
[11] *CMS picture* URL http://cylindricalonion.web.cern.ch/blog/201411/waking-sleeping-giant-cms-magnet-comes-back-life
[12] CMS 2010 *Physics Letters B* **692** 83 – 104 ISSN 0370-2693 URL http://www.sciencedirect.com/science/article/pii/S0370269310008725
[13] Baxendale J M, Hume C J and Thompson M G 1975 *Journal of Physics G: Nuclear Physics* **1** 781 URL http://stacks.iop.org/0305-4616/1/i=7/a=012
[14] Rastin B C 1984 *Journal of Physics G: Nuclear Physics* **10** 1609 URL http://stacks.iop.org/0305-4616/10/i=11/a=017
[15] Hebbeker T and Timmermans C 2002 *Astroparticle Physics* **18** 107 – 127 ISSN 0927-6505 URL http://www.sciencedirect.com/science/article/pii/S0927650501001803
[16] Collaboration L 2004 *Physics Letters B* **598** 15 – 32 ISSN 0370-2693 URL http://www.sciencedirect.com/science/article/pii/S0370269304011438
[17] ALICE 2016 *Journal of Cosmology and Astroparticle Physics* **2016** 032 URL http://stacks.iop.org/1475-7516/2016/i=01/a=032
[18] Cahuantzi M R 2015 *PoS* **ICRC2015** URL http://pos.sissa.it/archive/conferences/236/423/ICRC2015_423.pdf
[19] Antchev G *et al.* 2011 *Europhys. Lett.* **96** 21002 (*Preprint* 1110.1395)

Anisotropy in the Arrival Directions of Ultrahigh-Energy Cosmic Rays

Luis Villaseñor

Benemérita Universidad Autónoma de Puebla, Puebla, Mexico, on leave of absence from UMSNH, Morelia, Mexico

E-mail: `lvillasen@gmail.com`

Abstract. In this article we illustrate, in an interactive way, the analysis and visualization of anisotropy properties in the arrival directions of ultrahigh-energy cosmic rays detected by the Telescope Array and the Pierre Auger experiments by using data released by both collaborations. We describe the use of several programs that we have written in Python and Julia languages for this purpose. We also discuss the potential sources and analyse the effect of correcting the arrival directions to take into account the deflections of the cosmic rays by the magnetic field of our galaxy for one specific model of the galactic magnetic field under several assumptions about the composition of the primary cosmic rays.

1. Introduction

One of the most important results in the field of ultrahigh-energy cosmic rays (UHECRs) is the discovery of a hotspot in the arrival directions of the highest-energy cosmic rays detected with the Telescope Array (TA) experiment in the northern sky [1]. In this article we describe a number of programs that we have written in Python and Julia that can be used to reproduce these results by using the data that the Telescope Array Collaboration has made public. In their paper they use Li-Ma [2] significances, in this article we show that the use of P_{value} produces the same results.

Since these lectures were given at the VI School on Cosmic Rays and Astrophysics [3] our emphasis was on providing the students with a complete set computational tools required to reproduce these results and to eventually use these tools to make new progress on their own, this article illustrates the use of all the programs in detail. This software was available to everybody during the lectures.

The measurement of the statistical significance of the deviation from isotropy in the arrival directions of the highest-energy cosmic rays, and in particular, the calculation of penalty factors to account for the search for such deviations, can benefit from the use of high performance computing (HPC) on a cluster of CPUs and, for that reason, we provided an elementary introduction to Monte Carlo (MC) methods and the use of the Message Passing Interface (MPI) library [4]. In the lectures we made extensive use of the Jupyter Notebook as an interactive environment to work with software code, plots, data, websites, etc., and for that reason we also provided an elementary introduction to the use of the Jupyter Notebook [5], previously known as IPython Notebook.

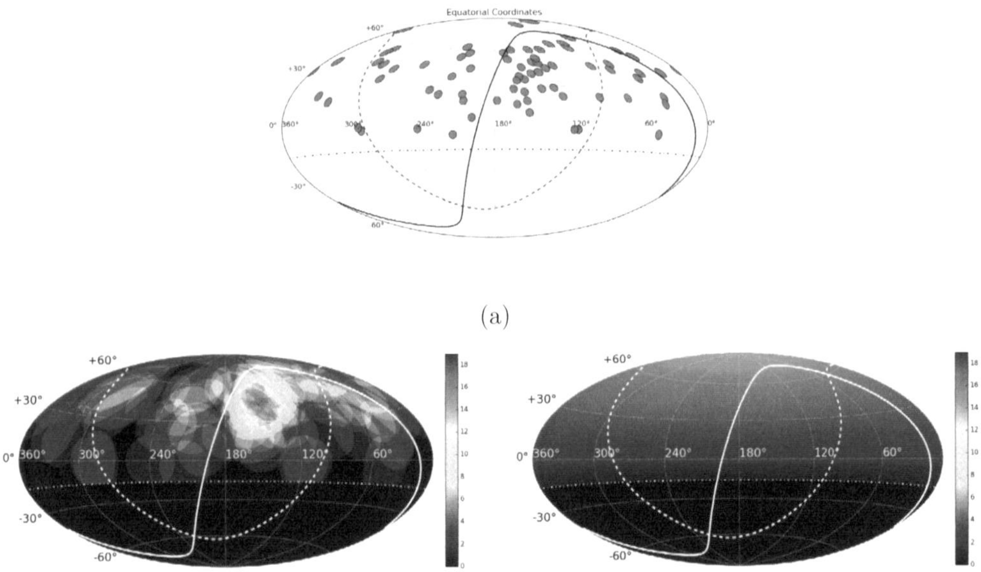

(a)

(b) (c)

Figure 1: Maps in equatorial coordinates using the Hammer projection for the data detected by the Telescope Array experiment [1]. The solid curve represents the supergalactic plane and the dashed line represents the galactic plane. The dotted line indicates the limit of the field of view of the TA experiment. (a) 72 UHECR events detected with $E >= 57$ EeV and zenith angle $< 55°$. (b) Same events oversampled with $R = 20°$. (c) One hundred thousand isotropic MC events generated in the field of view of the TA experiment, oversampled with $R = 20°$ and normalised to 72 events.

The same procedure and programs were used during the lectures to reproduce the anisotropy results in the arrival directions of the highest-energy cosmic rays in the southern sky [6] reported by the Pierre Auger Collaboration .

In addition, we describe the source code that we have written to backpropagate these cosmic rays, from the Earth to the edge of our galaxy, in the presence of galactic magnetic fields. We use the Julia language and apply it to a particular model of the galactic field for different assumptions about the composition of the primary cosmic rays.

The programs described in this article are presently available for download [7] and can be used freely by anyone, they have been tested in Mac OS X and Ubuntu operating systems. Finally, in order to facilitate the use of this article in an interactive way, the commands used to produce the results, and to create the plots, are shown in blue.

2. Hotspot Discovered by the Telescope Array Experiment

The hotspot in the arrival directions of ultrahigh-energy cosmic rays, discovered by the Telescope Array Collaboration in the northern sky, was reported in [1] for energies above 57 EeV with a statistical significance of 3.4σ. In this section we use the data released by the TA Collaboration in [1] to reproduce their results with our own analysis software.

2.1. Data Sample

The data for the arrival directions of the UHECRs detected by the TA experiment can be downloaded in a complete form from the electronic edition their paper. These data consist of 72 events with $E > 57$ EeV and zenith angle $\theta < 55°$ recorded from 2008 to 2013. The following data are provided for each event: date and time of occurrence, measured zenith angle, measured energy, and the equatorial coordinates of the arrival direction given by right ascension (RA or α) and declination (DEC or δ).

2.2. Data analysis

We will follow the analysis procedure described in [1]. First we copy the data and save it in a file named `TA_AstrophysJLett_2014.dat`. The format of this file can be displayed by the following command:

```
head -5 TA_AstrophysJLett_2014.dat
```

which produces the following output:

```
#Date and Time (UTC) Zenith(deg) E(EeV) RA(deg) DEC(deg)
2008 Jun 10 17:05:37 46.91 88.8  93.50 20.82
2008 Jun 25 19:45:52 31.98 82.6  68.86 19.20
2008 Jun 29 08:22:45 41.20 101.4 285.74 -1.69
2008 Jul 15 05:26:31 34.26 57.3  308.45 53.91
```

Next we change a few dashes for minus signs, convert from equatorial to galactic coordinates, apply some cuts and visualise the events with the program `PlotEvents.py`:

```
python PlotEvents.py TA
```

which creates the plot shown in Fig. 1.a.

In order to follow the same anisotropy analysis described in [1] we must now perform an oversampling with a radius of 20° around every point in the sky by using a grid of $0.1° \times 0.1°$. This time we use the Julia language to speed up the calculations and we optimise our code through the use of the k-dimensional tree algorithm[8]. The Julia program that we use for this purpose is named `OverSample.jl`; the command to oversample the TA events with a radius=20° is the following:

```
julia OverSample.jl TA 20
```

this program creates an oversampled map of the Telescope Array events and stores it in the file named `N_on.dat`. Next we use the program `PlotMap.py` to visualise this map with the command:

```
python PlotMap.py TA
```

which creates the plot shown in Fig. 1.b.

The next step consists in the evaluation of the background noise under the assumption that this noise is isotropic. For this purpose we generate 100 000 simulated arrival directions, taking into account the geometrical exposure as a function of zenith angle, θ, for the Telescope Array given by Eq. 1.

$$\frac{d\Omega}{dN} = N_\circ sin\theta cos\theta \tag{1}$$

The program that generates these Monte Carlo events is called `MCSim.jl`, we therefore execute the command:

```
julia MCSim.jl
```

IOP Conf. Series: Journal of Physics: Conf. Series **866** (2017) 012014 doi:10.1088/1742-6596/866/1/012014

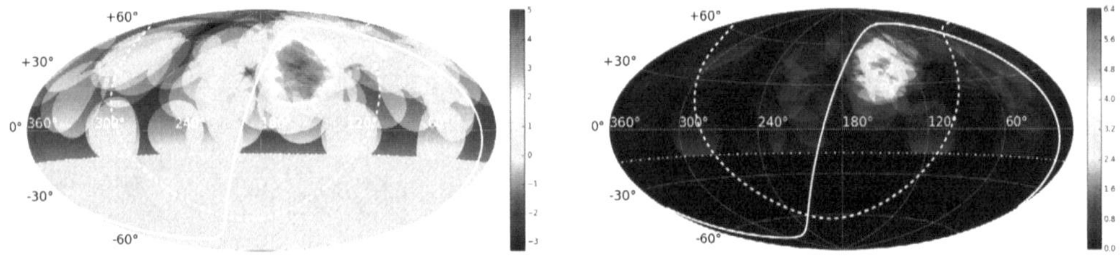

(a) (b)

Figure 2: Maps in equatorial coordinates using the Hammer projection for the data detected by the Telescope Array experiment [1]. The solid curve represents the supergalactic plane and the dashed line represents the galactic plane. The dotted line indicates the limit of the field of view of the TA experiment. (a) Map of the Li-Ma significance defined in Eq. 2. (b) Map of P_{value} defined in Eq. 3 and Eq. 4.

which creates a text file named `100K_evts.txt`. Next we apply the same oversampling procedure that we applied to the data and normalise to a total of 72 events. We therefore run again the program `OverSample.jl`, but this time applied to MC data simulated in the field of view of the TA experiment:

```
julia OverSample.jl Sim_TA 20
```

this program creates an oversampled map of the MC isotropic events generated in the field of view of the Telescope Array experiment with the same grid of 1800×3600 points and the same oversampling radius of 20° that we used for the data. The map generated is stored in the file named `N_off.dat`.

Since the number of events is higher, the execution of this program takes a few minutes in a typical laptop. One alternative to speed it up is to use MPI to parallelise the execution in case one can use a multiprocessor computing system.

In turn we use the program `PlotMap.py` to visualise this map with the command:

```
python PlotMap.py Sim_TA
```

that creates the plot shown in Fig. 1.c, the scale of its color bar is adjusted to match that of Fig. 1.b.

2.3. Statistical significance based on S_{LM}

Now we proceed to use both the oversampled data, `N_on.dat` shown in Fig. 1.b, and the oversampled isotropic background, `N_off.dat` shown in Fig. 1.c, to quantify the deviation from isotropy present in the data.

For this purpose we use the value of the Li-Ma significance [2], denoted by S_{LM}, calculated for each pixel by means of Eq. 2, where $\alpha = 72/100000$ is the ratio of the number of events in the data sample to the number of isotropic MC events.

$$S_{LM} = \left[2N_{on}ln\left(\frac{(1+1/\alpha)N_{on}}{N_{on}+N_{off}}\right) + 2N_{off}ln\left(\frac{(1+\alpha)N_{off}}{N_{on}+N_{off}}\right)\right]^{1/2} \quad (2)$$

We use the program named `LiMaSigma.jl` through the following command:

```
julia LiMaSigma.jl TA
```

this program produces the following output:

```
Max Val S_LM=5.04
Max Val -log10(P_VALUE)=6.45
```

and creates the files named `S_LM.dat` that in turn can be visualised with the `PlotMap.py` program as follows:

```
python PlotMap.py S_LM_TA
```

to generate the Li-Ma significance map shown in Fig. 2.a.

2.4. Statistical significance based on P_{value}

An alternative to the use of the Li-Ma significance is the use of P_{value} [9] to measure the deviation from isotropy. The P_{value} is defined in Eq. 3 and Eq. 4.

$$P_{value} = P_{Poisson}(X \geq N_{on}) \tag{3}$$

$$P_{value} = 1 - CDF_{Poisson}(\mu = \alpha N_{off}, X = N_{on} - 1) \tag{4}$$

where $CDF_{Poisson}$ is the Poisson cumulative distribution function.

The program `LiMaSigma.jl` that we run above also generated the P_{value} significance map `P_value.dat` which can be visualised with the following command:

```
python PlotMap.py P_value_TA
```

shown in Fig. 2.b.

In order to measure the statistical significance of the hotspot we now simulate a large number of sets of MC 72 events generated isotropically and we find out that only a fraction of 3.7×10^{-4} have a $\sigma_{LiMa} > 5.1$. We also find that the same fraction of sets have a $P_{value} < 10^{-6.4}$ meaning that using P_{value} is equivalent to using S_{LM}. These numbers correspond to a statistical significance of 3.4σ one-sided.

3. Hotspot Discovered by the Pierre Auger Observatory in the Southern Sky

The code we have used to reproduce the hotspot discovered with the Telescope Array experiment can be applied straightforwardly to the data released by the Pierre Auger Collaboration in 2015 [6] which also exhibits a hotspot located at $RA = 198.0°$ and DEC$=-25.2°$ in the southern sky, although with a lower statistical significance.

First we copy the data released in [6] and save it in a file named `Auger_AstrophysJ804_2015.dat`. The format of this file can be displaying by printing a few lines with the following command:

```
head -5 Auger_AstrophysJ804_2015.dat
```

which produces the following output:

```
#Year JulianDay Zenith(deg) E(EeV) RA(deg) DEC(deg) l(deg) b(deg)
2004 125 47.7 62.2  267.2  -11.4  15.5  8.4
2004 142 59.2 84.7  199.7  -34.9  -50.8  27.7
2004 177 71.5 54.6  12.7  -56.6  -56.9  -60.5
2004 239 58.3 54.0  32.7  -85.0  -59.1  -31.8
```

Next we change dashes for minus signs, apply some cuts and visualise the events with the program `PlotEvents.py`:

```
python PlotEvents.py Auger
```

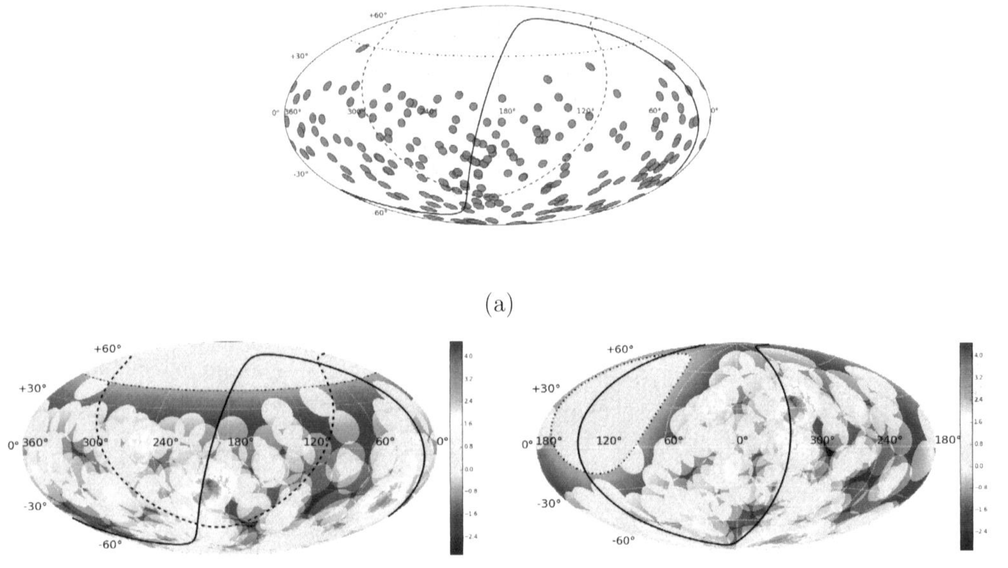

(a)

(b) (c)

Figure 3: Maps using the Hammer projection for the data realeased by the Pierre Auger Collaboration [6]. The solid curve represents the supergalactic plane and the dashed line represents the galactic plane. The dotted line indicates the limit of the field of view of the Pierre Auger Observatory. (a) Map in equatorial coordinates of the 200 UHECR events detected with $E >= 54$ EeV and zenith angle $< 80°$. (b) Map in equatorial coordinates of the Li-Ma significance defined in Eq. 2 for an oversampling radius of 12°. (c) Same as Fig. 3.b but in galactic coordinates.

which creates the plot shown in Fig. 3.a.

The same procedure and same programs as those described above can be used to measure σ_{LM} and P_{value}. Fig. 3.b shows the Li-Ma significance map in equatorial coordinates for an oversampling radius of 12° and Fig. 3.c shows the same Li-Ma significance map in galactic coordinates. The maximum value $\sigma_{LM} = 4.6$ occurs at $RA = 198.0°$ and $DEC = -25.2°$. Application of a penalty factor to account for the search procedure reduces the statistical significance as discussed in [6].

4. Potential Sources of Ultrahigh-Energy Cosmic Rays

In search of a correlation with the positions of potential astrophysical sources of the UHECRs we use the Swift BAT 70-Month Hard X-ray Survey, available from [10], which includes 1210 hard X-ray sources with the majority of them classified as AGN.

First we can copy these data with the command:

```
wget  http://swift.gsfc.nasa.gov/results/bs70mon/inc/data/BAT_70m_catalog_20nov2012.txt
```

To read these data we use:

```
python Read_BAT_Swift.py
```

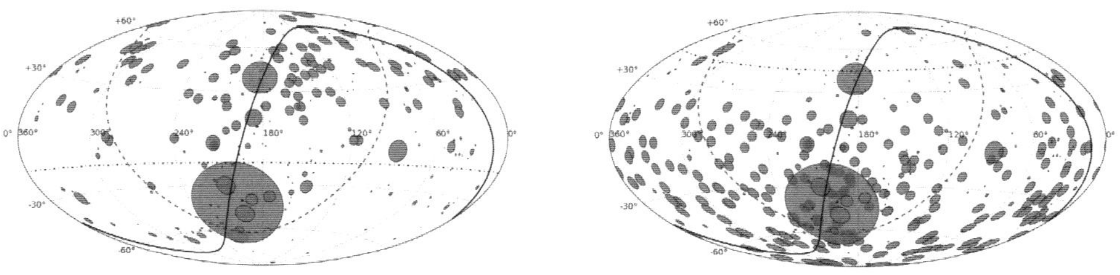

(a) (b)

Figure 4: Maps in equatorial coordinates using the Hammer projection of displaying the arrival directions of UHECR events and the positions of objects from the Swift BAT 70-Month Hard X-ray Survey. The solid curve represents the supergalactic plane and the dashed line represents the galactic plane. The red circles have a radius of 3° and represent the detected events. The green circles represent the 202 objects located within 100 Mpc from the Earth, their radii are arbitrarily scaled as $Flux/50$ in degrees. (a) Events detected by the Telescope Array experiment previously shown in Fig. 1.a . (b) Events detected by the Pierre Auger Observatory previously shown in Fig. 3.a.

which produces the following output with the flux listed in descending order for the 202 objects with $redshift < 0.0235$ (roughly equivalent to a $distance < 100$ Mpc) :

```
Total number of objects= 1210
Total number of objects with measured redshift < 0.0235= 202
Name  RA  DEC  Flux  Redshift
Cen A                          201.38  -43.03  1388.99  0.0018
NGC 4151                       182.67  39.42  538.93  0.0033
NGC 2110                       88.04  -7.43  319.13  0.0078
IC 4329A                       207.32  -30.33  290.49  0.016
NGC 4945                       196.34  -49.47  284.86  0.0019
NGC 4388                       186.45  12.68  277.38  0.0084
Circinus Galaxy                213.29  -65.33  272.12  0.0014
NGC 5506                       213.31  -3.24  241.03  0.0062
MCG -05-23-016                 146.93  -30.94  201.3  0.0085
4C 50.55                       321.18  50.95  200.46  0.02
NGC 4507                       188.87  -39.89  188.11  0.0118
NGC 3783                       174.76  -37.74  181.11  0.0097
NGC 7172                       330.5  -31.87  170.93  0.0087
Mrk 348                        12.22  31.96  156.04  0.015
Mrk 3                          93.84  71.04  140.96  0.0135
MCG +08-11-011                 88.76  46.45  133.44  0.0205
NGC 3516                       166.66  72.58  118.34  0.0088
AX J1737.4-2907                264.37  -29.15  115.86  0.0214
NGC 5252                       204.58  4.54  115.48  0.023
ESO 103-035                    279.56  -65.44  111.73  0.0133
 .......
2MASX J18305065+0928414        277.71  9.48  3.11  0.019
```

IOP Conf. Series: Journal of Physics: Conf. Series **866** (2017) 012014	doi:10.1088/1742-6596/866/1/012014

where RA and DEC are measured in degree, and the flux, measured by BAT-Swift in the 14-195 KeV band, is measured in units of 10^{-12} $ergs\ cm^{-2}s^{-1}$.

Next we proceed to make a map of these potential sources of UHECRs superimposed on the Telescope Array events by typing the following command:

```
python PlotEvents.py TA Cat
```

which produces the plot shown in Fig. 4.a.

Likewise, we can make a map of these potential sources superimposed on the Auger events with the command:

```
python PlotEvents.py Auger Cat
```

which creates the plot shown in Fig. 4.b.

The radii displayed for the Swift BAT 70-Month Hard X-ray Survey are arbitrarily scaled. In the absence of a model, we use weights that are directly proportional to the absolute luminosity measured in the 14-195 KeV band, in such a way that the D^2 factor in the absolute luminosity cancels out with the D^2 in the denominator due to the solid angle that our Galaxy presents as a target. Therefore we use $R = Flux/50$ in degrees with Flux in 10^{-12} $ergs\ cm^{-2}s^{-1}$ for the radii of these astrophysical objects, and $R = 3°$ for the radii of the events.

We arbitrarily limit the redshift of the plotted objects to < 0.0235 which corresponds to 100 Mpc when we convert with [11] . Note that using redshifts as a measure of distance is only a first approximation as it excludes possible motions of the objects within their local gravitational fields, a better approximation would be to use the measured distances available in [12].

Fig. 4 suggests that, assuming that the production of UHECRs is correlated with the production of hard X rays in the energy band measured by BAT-Swift (14-195 KeV), the hotspot observed in the northern sky can be explaining as dominated by two sources, namely NGC 4151 and NGC 4388, displayed in the northern sky of Fig. 4.a as green circles with radii 10.8° and 5.6°, respectively, whereas the southern hotspot, reported by Auger with a lower statistical significance, can be attributed to a few sources, namely CenA, NGC 2110, IC 4329A, NGC 4945, the Circinus Galaxy and NGC 5506, displayed in the southern sky of Fig. 4.b as green circles with radii equal to 27.8°, 6.4°, 5.8°, 5.7°, 5.4° and 4.8°, respectively. The fact that the northern sky has fewer potential sources in this context may explain its more compact hotspot with respect to the southern sky.

A similar exercise could be done using the Veron Catalog of Quasars & AGN, 13th Edition [13]. However this catalog is known to be statistically incomplete and non-uniform, although for sources within 100 Mpc the incompleteness and the non-uniformity may not represent a problem, another inconvenience of the Veron Catalog is the poor precision on the redshifts assigned to the AGN, although this can be easily fixed by using the more precisely measured distances available from [12].

5. Correction due to the Galactic and Extragalactic Magnetic Fields

The dream of doing astronomy with UHECRs, based on the assumption that the deflections of the cosmic rays at these extreme energies would be small enough to point to the sources, has gradually disappeared as we collect more data. This means that either the cosmic rays are light nuclei but the galactic and extragalactic fields are more intense than expected, or the composition is dominated by heavy nuclei, or both.

As we gradually discover more details about the intensities and structures of the galactic and extragalactic magnetic fields, and improve on the measurement of the composition of the primary cosmic rays, we will, in principle, apply better corrections that may eventually lead to the identification of the most important sources. In addition, the knowledge about the potential sources derived from multi-messenger approaches will lead to a better understanding of the acceleration mechanisms, as well as the ranges of energies and the nature of the primaries.

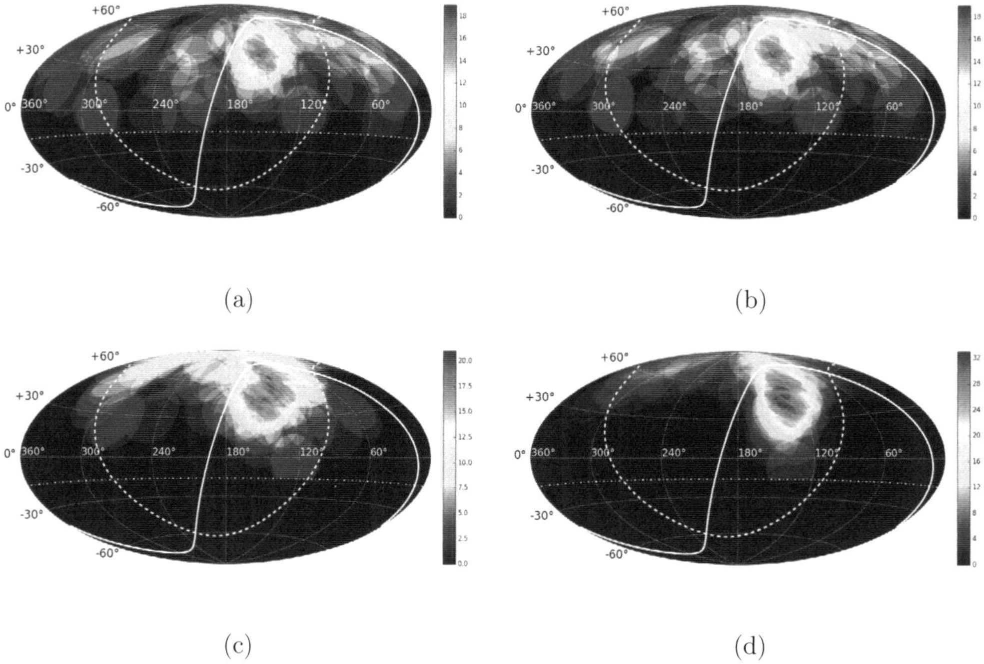

Figure 5: Maps in equatorial coordinates using the Hammer projection for the UHECR events detected by the Telescope Array experiment [1]. The solid curve represents the supergalactic plane and the dashed line represents the galactic plane. The dotted line indicates the limit of the field of view of the TA experiment. The oversampling radius is 20° in all plots. In (b), (c) and (d) the galactic magnetic field is approximated by the regular component of the JF2012 model. (a) Original events. (b) Events backpropagated to the edge of the Galaxy assuming pure proton composition. (c) Events backpropagated to the edge of the Galaxy assuming pure oxygen nuclei composition. (d) Events backpropagated to the edge of the Galaxy assuming pure iron nuclei composition.

In this section we describe the use of the required computational tools to correct for the deflections of the detected UHECRs in the presence of a particular model of the galactic magnetic field, called JF2012 and derived from fits to the WMAP7 Galactic Synchrotron Emission map and more than forty thousand extragalactic rotation measures [14, 15].

In particular we describe the use of a set of programs, written in Julia language and including Bash scripts, to backpropagate charged particles from the Earth to the edge of the Milky Way galaxy. We use the JF2012 model for the galactic magnetic field and use different assumptions about the composition of the UHECRs detected by the TA and the Pierre Auger collaborations.

The program named `CRBackProp.jl` uses the adaptive Cash-Karp method [16] to solve the Lorentz force equation to obtain the trajectories of the backpropagated UHECRs and the program `Propa_JF2012.jl` translates the JF2012 model into Julia code. Other alternative programs to backpropagate cosmic rays are [17] and [18].

As an example, if we want to backpropagate an iron nuclei that arrives with an energy of 60 EeV from the direction *galactic longitude* $= 40°$ and *galactic latitude* $= -20°$, we run the Julia program named `CRBack_Prop_example.jl` listed here:

```
include("JF2012.jl")
```

IOP Conf. Series: Journal of Physics: Conf. Series **866** (2017) 012014 doi:10.1088/1742-6596/866/1/012014

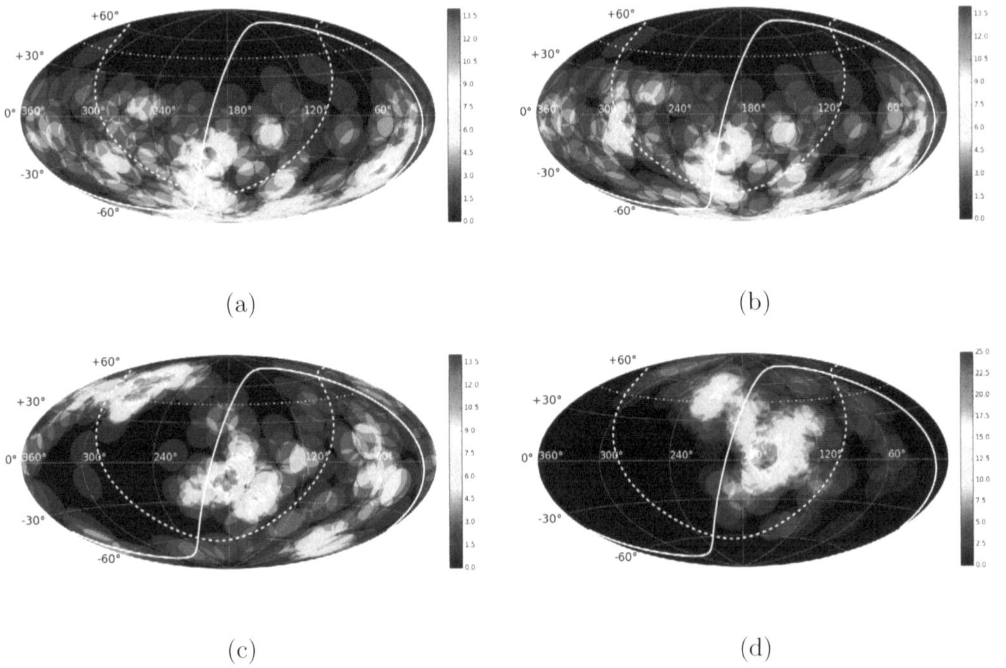

Figure 6: Maps in equatorial coordinates using the Hammer projection for the UHECR events detected by the Pierre Auger Observatory [6]. The solid curve represents the supergalactic plane and the dashed line represents the galactic plane. The dotted line indicates the limit of the field of view of the Auger experiment. The oversampling radius is 12° in all plots. In (b), (c) and (d) the galactic magnetic field is approximated by the regular component of the JF2012 model. (a) Original events. (b) Events backpropagated to the edge of the Galaxy assuming pure proton composition. (c) Events backpropagated to the edge of the Galaxy assuming pure oxygen nuclei composition. (d) Events backpropagated to the edge of the Galaxy assuming pure iron nuclei composition.

```
include("Cash-Karp.jl")
include("Propa_JF2012.jl")
alpha = Propa_JF2012(26,56,60.,40.,-20.,1,true,false,false,20000.)
```

This program produces the following output:

```
niter   niter_advance  E_f x_f y_f z_f l_f  b_f  alpha  t_f
101   94  60.01 1326.01 19344.98 6714.38 86.07 19.09 59.72 66980.37
```

returning the number of iterations, the energy (in EeV), the positions of the particle (in pc), the galactic longitude and galactic latitude of the entrance direction at the edge of the Galaxy (in degrees), the deflection between the entrance direction at the edge of the Galaxy and the arrival direction at the Earth (in degrees) and the time to transit our galaxy (in years).

In order to backpropagate all the events detected by the TA experiment we use the Bash script named `Backprop.sh`. Assuming all the detected events are protons we backpropagate them by using the following sequence of commands:

```
python PlotEvents.py TA
sh Backprop.sh p
```

IOP Conf. Series: Journal of Physics: Conf. Series **866** (2017) 012014 doi:10.1088/1742-6596/866/1/012014

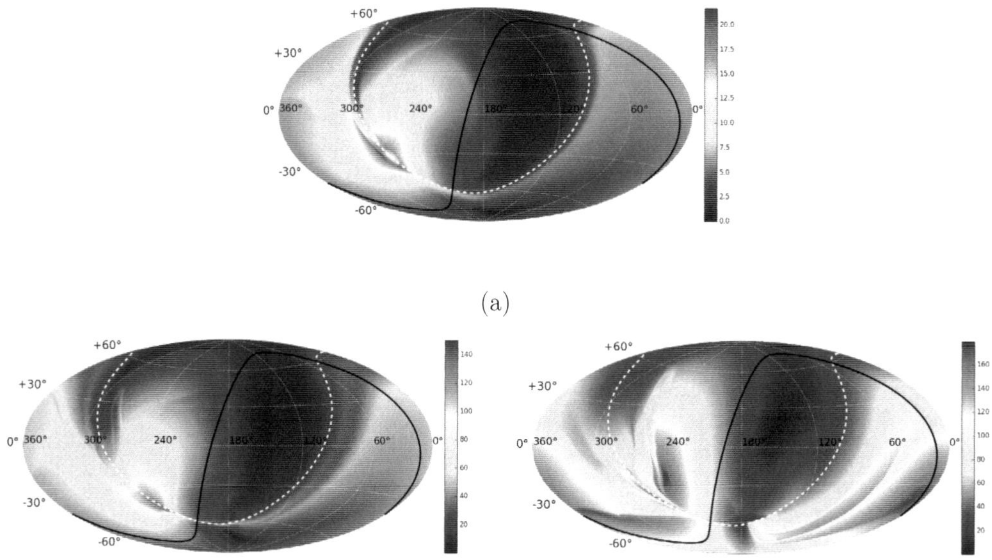

(a)

(b) (c)

Figure 7: Deflection angles in degrees for particles arriving at each point of these maps with E = 60 EeV. These particles are backpropagated to the edge of the Galaxy with the galactic magnetic field approximated by the regular component of the JF2012 model. All the maps are in equatorial coordinates using the Hammer projection. The solid curve represents the supergalactic plane and the dashed line represents the galactic plane. (a) Assuming the particles are protons. (b) Assuming the particles are oxygen nuclei. (c) Assuming the particles are iron nuclei.

which produces a file named `Data.p.jf2012.dat`. Next we use the following sequence of commands:

```
python PlotEvents.py TA p
julia Oversample.jl TA 20
```

to oversample with a radius = 20°, and finally the command:

```
python PlotMap.py TA
```

to create a plot with the oversampled map for the backpropagated events. If we backpropagated these events under the assumption that they are oxygen or iron nuclei we simply replace p by O or Fe, respectively, in the above sequence of commands.

The procedure described gives rise to the plots shown in Fig. 5 under the assumptions that all the UHECRs are protons, oxygen nuclei or iron nuclei, respectively. Note that we have used only the so-called regular component of the field in the JF2012 model, but one can activate the striated and/or the turbulent component of the model as well. This is done by changing one or the two *false* variables to *true* in the call to the function `Propa_JF2012`.

In a similar way we obtain the oversampled maps for the backpropagated events detected by the Pierre Auger Observatory, according to [6] this time we use an oversampling radius = 12°; to do this we just replace TA by Auger in the sequence of commands given above. The results are shown in Fig. 6.

IOP Conf. Series: Journal of Physics: Conf. Series **866** (2017) 012014 doi:10.1088/1742-6596/866/1/012014

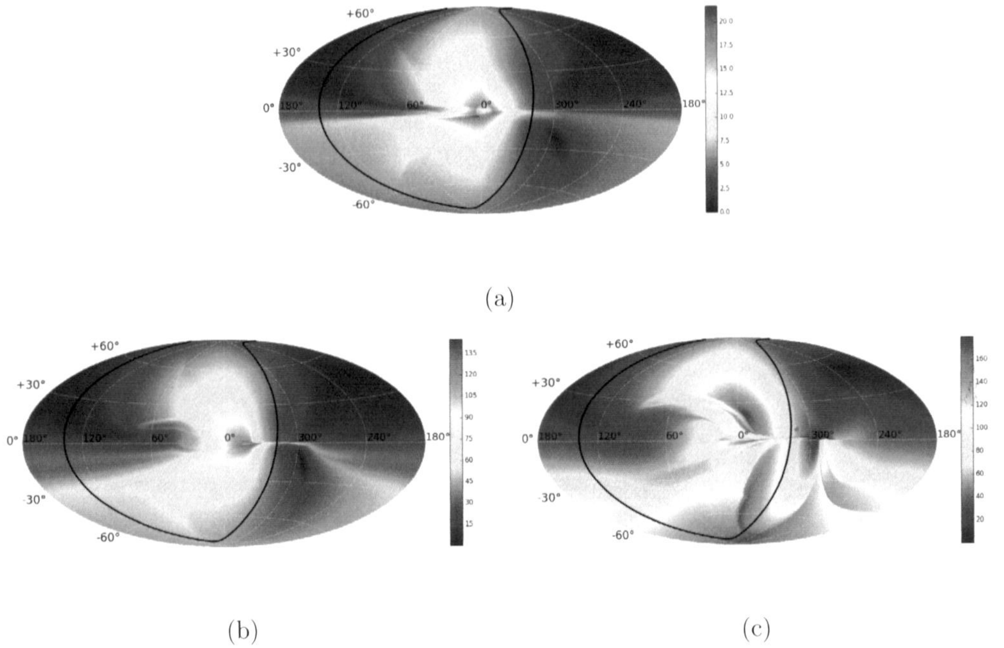

Figure 8: Deflection angles in degrees for particles arriving at each point of these maps with E = 60 EeV. These particles are backpropagated to the edge of the Galaxy with the galactic magnetic field approximated by the regular component of the JF2012 model. All the maps are in galactic coordinates using the Hammer projection. The solid curve represents the supergalactic plane. (a) Assuming the particles are protons. (b) Assuming the particles are oxygen nuclei, (c) Assuming the particles are iron nuclei.

We can conclude that both hotspots, in the northern and southern sky, are magnified if we take into account the deflections produced by the galactic magnetic field, approximated by the JF2012 model, under the assumption of a composition dominated by medium or heavy nuclei. The positions of the hotspots are near the positions of the uncorrected hotspots, and both are on the same side of the supergalactic plane. We are in the process of adding the deflections due to the extragalactic magnetic model using some of the proposed models for the latter.

Above we suggested the idea that the lower compactness of the southern hotspot may be due to the contribution of more sources from the Swift BAT 70-Month Hard X-ray Survey with respect to the northern sky, this idea may be reinforced if the different potential sources accelerate predominantly different particles at these extreme energies leading to more disperse angular deflections due to the galactic and extragalactic magnetic fields.

It is evident that a more quantitative correlation study is required to reinforce or discard some of these ideas. Likewise, it will be very interesting so study the effect in the positions and compactness of the hotspots due to the inclusion of deflections caused by extragalactic magnetic fields.

6. Maps of Deflection Angles due to the Galactic Magnetic Field for the JF2012 Model

By using the same Julia programs described above we can backpropagate cosmic rays of a fixed energy for different compositions. To do this, we divide the full sky in a grid of $180 \times 5 \times 360 \times 5$ points, i.e., we use the *granularity* variable equal to 5 instead of 10 as we previously did, and backpropagate protons, oxygen nuclei and iron nuclei arriving at the points of this grid with an energy of 60 EeV. The back propagation to the edge of the Galaxy assumes a galactic magnetic field approximated by the regular component of the JF2012 model.

Given that we need to backpropagate over 1.6 million particles, and since their trajectories are independent, we use MPI [4] in a multiprocessor HPC cluster to speed up the execution of this task which is trivially parallelizable. We use the following commands:

```
mpirun -n N -machinefile hosts julia mpi_map.jl EQ
```

and

```
mpirun -n N -machinefile hosts julia mpi_map.jl Gal
```

to obtain deflection maps in equatorial and galactic coordinates, respectively. The file `hosts` contains the names of the nodes of your cluster to use a total of N cores, where N can be small or large depending on your computing resources.

Fig. 7 shows the maps of deflections for protons, oxygen nuclei and iron nuclei of 60 EeV backpropagated in equatorial coordinates and Fig. 8 shows the same in galactic coordinates.

7. Conclusions

In this article we have described the use of a number of programs that we have written in Python and Julia that can be used to reproduce data analyses that lead to the discovery of a hotspot in the arrival directions of the highest-energy cosmic rays detected with the Telescope Array experiment in the northern sky, and also to reproduce the findings of the Pierre Auger Collaboration in the southern sky. We do this by using data released by both collaborations.

We have also described the use of programs that can be used to correlate the arrival directions of the detected events with the locations in the sky of possible sources from the Swift BAT 70-Month Hard X-ray Survey.

We have also presented the analysis of the effect of correcting the arrival directions, reported by both experiments, to take into account the deflections of the cosmic rays caused by the magnetic field of our galaxy. We have described the use of these programs for one specific model of the galactic magnetic field under several assumptions about the composition of the primary cosmic rays.

In particular we have shown that both hotspots, in the northern and southern skys, are magnified if we take into account the deflections produced by the galactic magnetic field, approximated by the JF2012 model, under the assumption of a composition dominated by medium to heavy nuclei. The positions of the corrected hotspots are near the positions of the uncorrected hotspots and both are on the same side of the supergalactic plane. It will be interesting to add deflections due to extragalactic magnetic fields using some of the proposed models for the latter that may eventually move the hotspots closer to the supergalactic plane, and closer to the potential sources mentioned above, namely CenA, NGC 2110, IC 4329A, NGC 4945, the Circinus Galaxy and NGC 5506 in the southern sky and NGC 4151 and NGC 4388 in the northern sky.

Finally, we have presented maps, in equatorial and galactic coordinates, of deflections of protons, oxygen nuclei and iron nuclei, of a fixed energy equal to 60 EeV, when they are backpropagated to the edge of our galaxy in a galactic magnetic field approximated by the regular component of the JF2012 model.

The programs that we have written for the analyses described in this article can be used freely by anybody, they can be downloaded from [7].

Appendix A: Installation of Python Modules

The use of Python may become unstable due to the accidental use of conflicting libraries, for this reason it is recommended to install a virtual environment though the use of *virtualenv* [19] which is a tool to create isolated Python environments. A simpler solution is to install the Python distribution from *anaconda* [20] and then type

```
conda update conda
conda create -p  CR python matplotlib basemap ipython
```

next you can activate the CR Python environment with the command:

```
source activate /YOUR-PATH/CR
```

The next step is to install the specific python modules required by using the *pip* command associated to the active virtual environment by means of the following commands:

```
pip install astropysics
pip install CosmoloPy
pip install pandas
```

next you can download astroconvert from [21] or use the following command:

```
wget http://vanderbiltastro.pbworks.com/w/file/fetch/46335397/astroconvert.py
```

If you want to go back to use your normal Python environment you just deactivate the CR environment with the command:

```
source deactivate /YOUR-PATH/CR
```

Appendix B: Installation of the Julia Language and Required Packages

The Julia language is a relatively new high-level, high-performance dynamic programming language created specifically for technical computing. It can be installed by downloading the appropriate file and following the platform specific instructions from [22]. We recommend to use version 0.4.2 to assure compatibility with the KDTrees package. If you want to use the MPI package it is necessary to install first the OpenMP application programming interface [23], check the installation instructions specific to your platform. The required packages to run the described programs in Julia can be installed by typyng the commands:

```
julia
Pkg.add("KDTrees")
Pkg.add("SkyCoords")
Pkg.add("MPI")
```

Acknowledgements

The author thanks CONACyT for making possible the creation of the Center for High Performance Computing of UMSNH in Morelia, and the Laboratorio Nacional de Supercómputo del Sureste de México, located in the BUAP Campus in Puebla; the Monte Carlo simulations of this article where performed in these facilities. The author also thanks CONACyT for the support during his leave of absence from UMSNH and for support received through the Mesoamerican Centre for Theoretically Physics located in Tuxtla Gutierrez, Chiapas, Mexico.

References

[1] R. U. Abbasi et al. [Telescope Array Collaboration], Astrophys. J. Lett. 790, L21 (2014)
[2] T. Li and Y. Ma, ApJ, 272 (1983) 317
[3] http://mctp.mx/e_VI_School_on_Cosmic_Rays_and_Astrophysics.html
[4] http://dl.acm.org/citation.cfm?id=898758
[5] https://ipython.org/notebook.html
[6] A. Aab et al. [Pierre Auger Collaboration], Astrophys.J. 804 (2015) 1, 15
[7] https://github.com/lvillasen/ToolsCosmicRays
[8] Bentley, J. L., "Multidimensional binary search trees used for associative searching". Communications of the ACM. 18 (1975) 9.
[9] Bhattacharya, Bhaskar; Habtzghi, DeSale (2002). "Median of the p value under the alternative hypothesis". The American Statistician.
[10] http://swift.gsfc.nasa.gov/results/bs70mon/
[11] http://roban.github.io/CosmoloPy/
[12] NASA/IPAC Extragalactic database, http://ned.ned.caltech.edu/forms/byname.html
[13] http://heasarc.gsfc.nasa.gov/W3Browse/galaxy-catalog/veroncat.html
[14] R. Jansson and G. R. Farrar, A New Model of the Galactic Magnetic Field, ApJ 757 (2012) 14.
[15] R. Jansson and G. R. Farrar, The Galactic Magnetic Field, Astrophys. J. 761 (2012) L11.
[16] J. R. Cash, A. H. Karp. "A variable order Runge-Kutta method for initial value problems with rapidly varying right-hand sides", Transactions on Mathematical Software 16: 201-222, 1990.
[17] M.S. Sutherland, B.M. Baughman, J.J. Beatty, Astroparticle Physics, Volume 34, Issue 4, p. 198-204. (2010)
[18] R.A. Batista, M. Erdmann, C. Evoli, K.-H. Kampert, D. Kuempel, G. Müller, G. Sigl, A. van Vliet, D. Walz, T. Winchen, EPJ Web of Conferences 99 (2015) 13004
[19] https://virtualenv.pypa.io/en/stable/
[20] https://www.continuum.io/downloads
[21] http://vanderbiltastro.pbworks.com/w/file/46335397/astroconvert.py
[22] http://julialang.org/downloads/
[23] http://openmp.org/wp/

IOP Conf. Series: Journal of Physics: Conf. Series **866** (2017) 012015 doi:10.1088/1742-6596/866/1/012015

High Performance Computing at the Universidad Autónoma de Chiapas

K. Caballero Mora, S. Estrada Jiménez, C. Franco Espinosa, J. Romo González and L. Valencia Palomo

Universidad Autónoma de Chiapas

E-mail: `lizardo.valencia@unach.mx`

Abstract. In these proceedings we report on the events that took place before, during and after the donation of 364 server that will be used to develop the High Peformance Computing at the Universidad Autnoma de Chiapas.

1. High Performance Computing

Nowadays the High Performance Computing (HPC) is a very important tool for the development, not only for science but also for our modern society. In the specific case of an university it becomes invaluable to provide access to the HPC to the academic, research and student community. The range of applications is quite wide, as it can vary from basic science to administrative procedures.

Chiapas, Central America and the Caribbean have necessity of HPC, unfortunately as they don't have this infrastructure, they have to perform their scientific calculations abroad. However, depending on an institution that can be far apart is a lack of growing from the point of view of the demanding part. In this sense, requesting permission to use an external facility is subject to the availability and priorities.

For this reason the European Organization for Nuclear Research (CERN) has decided to donate an important amount of computer equipment to the Universidad Autnoma de Chiapas (UNACH). The donation consists of 364 servers with a commercial cost of, approximately, 2 million dollars.

2. European Organization for Nuclear Research

The European Organization for Nuclear Research has its origins in the mid twentieth century. Right after World War II, the european countries noticed the importance of scientific research on Nuclear Physics. So, Pierre Auger, Louis de Broglie and Niels Bohr, together with some other colleagues pushed for the construction of a common european facility focused in this topic. The idea was to encourage the exchange of scientific knowledge but also to share the high costs of the construction and maintenance.

Along his more than 60 years of history, CERN has conducted an important amount of scientific research in the fields of Nuclear and Particle Physics. In 1965, at the same time as a laboratory in the US, they detected the first anti-nuclei conformed by an anti-proton and an anti-neutron. In 1983, by colliding proton and anti-proton beams, UA1 and UA2 experiments led the detection of W and Z bosons. One year later the leading scientists, Simon van der Meer and Carlo Rubbia were awarded with the Nobel Physics Prize. On the 4th of July of 2012, ATLAS and CMS experiments at the Large Hadron Collider (LHC) announced the discovery of the Brout - Englert - Higgs Boson. The very next year the Nobel Physics Prize was awarded to Peter Higgs and Francois Englert.

The CERN is composed of 21 "state members", most of them being europeans. There are also 6 other countries with the status of "associated members" and 5 more countries catalogued as "observers". Finally CERN has also collaboration agreements with 35 countries, among them Mexico.

With the development of the new facilities, the experiments started to become more complex and also more expensive. By the 1980's CERN began to search for collaborators outside Europe. By the end of this decade, in order to motivate the exchange of scientific data, the physicist Tim Berners-Lee created the World Wide Web. More than 25 years after Berners-Lee's development, due to the requirements imposed by the LHC, the informatic services have been improved. At the LHC there are around 600 million proton-proton collisions per second at energies never achieved by humankind before. These collisions are registered by 150 million sensors, resulting in an output of 700 megabytes of data per second, this is, one CD of information every second. By the end of one year of operations at the LHC, we would have approximately 30 million Gigabytes of information.

It is clear that there is no single computer nor a single data center that can store and process such amount of information. So, in order to overcome this obstacle the World Wide LHC Computing Grid was created. This is a world network of around 170 supercomputing centres from 42 countries all around the world. In Mexico only the UNAM is part of the Grid.

3. High Performance Computing Laboratory

The history of the donation of the servers can be traced back to 2014 when Dr. Luis Flores, from Hong Kong University offered to the UNACH to request a donation to CERN. The initial team of academic personnel at the UNACH working on this process was: Dr. Karen Caballero Dr. Sendic Estrada, Dr. Carolina Franco, LIA Jorge Romo and Professor Arnulfo Zepeda. This team started some negotiations with the corresponding authorities of the UNACH, the main problematics that arose during those meetings were related to the place to store the servers once they reached the university and the transportation of all the material (around 13 tons).

Indeed, the donation consisted of equipment that was previously used at CERN Data Center. However, when the computers can no longer satisfy the extremely high requirements of CERN, they are replaced by new and more powerful computers. However, the old computers are still useful and can be safely used in universities or scientific research institutions that do not have so high computing requirements.

The 364 servers, 24 switches and 26 racks were shipped from CERN, on August 26, with Tuxtla Gutirrez as destination. Finally, the equipment arrived on October 11th to the UNACH. The arrival of computers from a top scientific research center to Chiapas was a mediatic phenomena, not only in the region but in all Mexico. Many reports and interviews were performed, local media spread the news during the following days and so the starting scientific activities in Chiapas were highlighted.

With the arrival of all this equipment it has been possible to create the High Performance Computing Laboratory (LARCAD) at the UNACH. The LARCAD is still under construction and it is expected to be finished by the end of the year. The main goals of this laboratory will be:

- Provide high efficient computing services to the academic and research staff of the UNACH.
- Obtain, as soon as possible, the category of Tier 1. In a medium term, the LARCAD should be a Tier 2 center.

Figure 1. *Left:* First meeting between academic and authorities from UNACH. *Right:* Shipment of the donation (August 16th, 2014).

Figure 2. *Left:* Arrival of the servers, racks and switches (October 11th, 2014). *Right:* Preliminary status of the LARCAD.

- Create human resources with high specialization in the performance of the laboratory.
- Incorporate more regional universities and institutions to the development of the HPC in the south of Mexico, Central America and the Caribbean.

Finally, although the original use of the servers was exclusively in the High Energy Physics field, they can be used in any other branch of science. In particular, the main topics that are being developed at the UNACH by the academic staff are:

- Astroparticle physics: in this field there are currently a couple of professors working in international collaboration as HAWC, Pierre Auger, LAGO and JEM-EUSO.
- Optics: lasers and optic fibers.
- Complex systems: non-linear dynamics and biologic processes.
- Theoretical physics: properties of Bose-Einstein condensates, thermodynamic properties of graphene, field theories, gravitation.
- Mathematics: pure and applied mathematics, topology, differential geometry, bio-mathematics.

- Energy: solar, eolian, hidro.
- Environment: climate change, ecology, sustainability.

4. Conlusion

In conclusion, the donation of computing servers from CERN will benefit not only to the UNACH but to all the south of Mexico. Currently the efforts are focused on the construction of the LARCAD and the first tests of the equipment. Is is expected that this data center is completed by the end of the present year.

References

[1] http://serversdonations.web.cern.ch/content/list-past-donations-servers-retired-cern-data-centre
[2] http://cds.cern.ch/record/2048937?ln=es
[3] http://dcs.unach.mx/index.php/noticia/item/3209-arriban-a-la-unach-los-servidores-inform%C3%A1ticos-donados-por-la-organizaci%C3%B3n-europea-para-la-investigaci%C3%B3n-nuclear.html
[4] J. Krige, History of CERN, Vol. 3.

The Dark Universe and the detection of Dark Matter

P. Belli

INFN, sez. Roma "Tor Vergata", I-00133 Rome, Italy

E-mail: pierluigi.belli@roma2.infn.it

Abstract.
Many experimental observations and theoretical arguments have pointed out that a large fraction of the Universe is composed by Dark Matter particles. Many possibilities are open on the nature and interaction types of such relic particles. Moreover, the poor knowledge of many fundamental astrophysical, nuclear and particle Physics aspects and of some experimental and theoretical parameters, the different used approaches and target materials, etc. make challenging the comparisons and the implications of the different experimental efforts. Some general arguments are addressed here and future perspectives are mentioned. A particular care will be given to the results obtained by exploiting the model independent Dark Matter annual modulation signature for the presence of Dark Matter particles in the galactic halo by DAMA experiment. Results from the other experiments using different procedures, different techniques and different target-materials will be shortly addressed as well as implications and experimental perspectives.

1. Introduction

Experimental observations and theoretical arguments have pointed out that most of the matter in the Universe has a non baryonic nature and is in form of Dark Matter (DM) particles. Many candidates – different in nature and with different and various interaction types – as DM particles of the Universe have been proposed within theories beyond the Standard Model of particle physics.

Depending on the DM candidate, the interaction processes can be various, as e.g.: 1) elastic scatterings on target nuclei with either spin-independent or spin-dependent or mixed coupling; moreover, an additional electromagnetic contribution can arise, in case of few GeV candidates, from the excitation of bound electrons by the recoiling nuclei [1]; 2) inelastic scatterings on target nuclei with either spin-independent or spin-dependent or mixed coupling in various scenarios [2, 3, 4, 5, 6, 7]; 3) interaction of light DM either on electrons or on nuclei with production of a lighter particle [8]; 4) preferred interaction with electrons [9]; 5) conversion of DM particles into electromagnetic radiation [10]; 6) etc.. Often, the elastic scattering on target nuclei is the considered interaction process, but other processes are possible and considered in literature, as those aforementioned where also electromagnetic radiation is produced. Hence, considering the richness of particle possibilities and the existing uncertainties on related astrophysical (e.g. halo model and related parameters, etc.), nuclear (e.g. form factors, spin factors, scaling laws, etc.) and particle physics (e.g. particle nature, interaction types, etc.), a widely-sensitive model independent approach is mandatory. Most activities in the field are instead based on a particular *a priori* assumption on the nature of the DM particle and of its interaction, in order to try to overcome the limitation arising from their generally large originally measured counting rate.

Moreover, many other experimental and theoretical uncertainties exist and must be properly considered in a suitable interpretation and comparison among experiments of direct detection of DM particles.

Before summarizing the efforts of the most important direct detection experiments, let us briefly comment few items. Firstly, the DM indirect search – that is the study of possible products either of decay or of annihilation of DM particles in the galactic halo or in celestial body – is performed as by-product of experiments located underground, under-water, under-ice, or in space. The interpretation of such a study is strongly dependent on the chosen assumptions for the modeling of the background and is restricted to some DM candidates with peculiar features. Therefore, all that shows the intrinsic uncertainties of the DM indirect searches to unambiguously assess presence of DM in the galactic halo. On the other hand, experiments at accelerators may prove – when they can state a solid model independent result – the existence of some possible DM candidate particles, but they could never credit by themselves that a certain particle is a/the only solution for DM particle(s). Moreover, DM candidate particles and scenarios (even e.g. in the case of the neutralino candidate) exist which cannot be investigated at accelerators.

2. The Dark Matter particles detection

In order to pursue a widely sensitive direct detection of DM particles in the galactic halo, a model independent approach, a ultra-low-background suitable target material, a very large exposure and the full control of running conditions are strictly necessary.

Indeed, most activities in the field release marginal exposures even after many years underground; they do not offer suitable information e.g. about operational stability and procedures during the running periods, and generally base their analysis on a particular *a priori* assumption on the nature of the DM particle and its interaction, and of all the used parameters.

In particular, the applied rejection and subtraction procedures to reduce the experimental counting rate, in order to derive a set of recoil-like candidates, is pursued by experiments as CDMS, EDELWEISS, CRESST, XENON, LUX, etc.. It is worth noting that the applied subtraction procedures are statistical and cannot offer an unambiguous identification of the presence of DM particle elastic scatterings because of the known existing recoil-like indistinguishable background; tails of the subtracted populations can play a role as well. Finally, the electromagnetic component of the counting rate, statistically "rejected" in this approach, can contain either the signal or part of it, and it will be lost. In the following few main experimental activities are mentioned as examples. Some arguments can be also found in Ref. [11, 12].

In the double read-out bolometric technique, the heat signal and the ionization signal are used in order to discriminate between electromagnetic and recoil-like events. This technique is used by CDMS and EDELWEISS collaborations. In particular, the CDMS-II detector consisted of 19 Ge bolometers of about 230 g each one and of 11 Si bolometers of about 100 g each one. The experiment released data for an exposure of about 194 kg $\times$ day using only 10 Ge detectors in the analysis (discarding all the data collected with the other ones) and considering selected time periods for each detector [13]. EDELWEISS employed a target fiducial mass of about 2 kg of Ge and has released data for an exposure of 384 kg $\times$ day collected in two different periods (July-Nov 08 and April 09-May 10) with a 17% reduction of exposure due to run selection [14]. These two experiments claim an "event by event" discrimination between noise + electromagnetic background and recoil + recoil-like (neutrons, end-range alphas, fission fragments,...) events by comparing the bolometer and the ionizing signals for each event. Thus, their results are, actually, largely based on huge data selections, as for example, the time cut analysis used to remove the so-called surface electrons that are distributed in both the electromagnetic and recoil bands. The stability, the nonlinear response and the robustness of the reconstruction procedure are key points, as well as the associated systematical errors. In these experiments few recoil-like events survive the many selections/subtractions cuts applied in the data analysis; these events are generally interpreted in terms of background. As regards, in particular, their application to the search for time dependence of the data (such as the annual modulation signature), it would require – among other – to face the objective difficulty to control all the operating conditions – at the needed level ($< 1\%$) – despite of the required periodical procedures e.g. for cooling and for calibration and owing to the limitation arising from the low duty cycle. For example, the attempt by CDMS-II to search for annual modulation in Ge target has been performed with a marginal exposure by using only 8 detectors over 30 and using – among others – data that are not continuous over the whole annual periods considered in the analysis [15] the use of non-overlapping time periods, collected with detectors having different background rate within the signal box does not allow one to get any reliable result in the investigation of an effect at few percent level (see e.g. arguments in Ref. [16]).

Other data taking was dedicated to measurements using a calorimetric technique, named CDMSlite, which relies on voltage-assisted Luke-Neganov amplification of the ionization energy deposited by particle interactions [17]. The data were collected with a single 0.6 kg germanium detector running for ten live days at the Soudan Underground Laboratory. A low energy threshold of 170 eV_{ee} (electron equivalent) was claimed [17], while recent data taking achieved even lower energy threshold [18]. In the meanwhile SuperCDMS at SNOlab [19] reported preliminary results corresponding to an exposure of 577 kg $\times$ days with an increased mass of 9.0 kg (15 detectors of 600 g each) and with increased detectors' performances [20]. Eleven events were observed not fully compatible with background expectation, even assuming the correctness of all the adopted procedures [20].

The results of CDMS-II with the Si detectors were published in two close-in-time data releases [21, 22]; while no events in six detectors (corresponding exposure of only 55.9 kg$\times$day before analysis cuts) were reported in the former [21], three events in eight detectors (corresponding raw exposure of 140.2 kg$\times$day) were reported over the residual background, estimated after subtraction: $\simeq 0.4$ in the second one [22]. The latter result could be interpreted – under certain assumptions – in terms of a DM candidate with spin-independent interaction and a mass around 10 GeV, which is compatible with some interpretations of the model independent DM annual modulation result already reported by DAMA in terms of this kind of DM candidate and with some other hints reported by CoGeNT (see later).

In the meanwhile EDELWEISS was in data taking in the period July 2014 – April 2015 and restarted in June 2015 with 36 detectors installed corresponding to a target fiducial mass of more than 14 kg of Ge [23]; new results collected with eight FID (Full Inter Digitized) detectors (582 kg $\times$ day) have been recently presented [24].

The CRESST experiment exploits the double read-out bolometric technique, using the heat signal due to an interacting particle in the $CaWO_4$ crystals and the produced scintillation light. A statistical discrimination of nuclear recoil-like events from electromagnetic radiation is performed. As regards the applied cuts and selection procedures, most of the above discussion still holds. A previous run (8 detectors of 300 g each one, for an exposure of about 730 kg $\times$ day) showed that, after selections, 67 nuclear recoil-like events were observed in the Oxygen band [25]. The background contribution estimated by authors ranges from 37 to 43 events, and does not account for all the observed events. The remaining number of events and their energy distribution could be interpreted – under certain assumptions – in terms of a DM candidate with spin-independent interaction and a mass in the range of 10-30 GeV. This result has been not confirmed in the last run [26], where a more marginal exposure has been used (52 kg $\times$ day and energy threshold of 307 eV), confirming the difficulties to manage the systematics in such experiments.

The new version of CRESST (CRESST-III) will use new detector modules of 24 g each trying to attain low energy thresholds. Projects of large mass bolometers are also planned in Europe

(EURECA) and at SNOlab.

The XENON project uses instead dual phase liquid/gas detectors. Experiments exploiting such technique (as also LUX, DARKSIDE) perform statistical discrimination between nuclear recoil-like candidates and electromagnetic component of the measured counting rate through the ratio of the prompt scintillation signal ($S1$) and the delayed signal ($S2$) due to drifted electrons in the gaseous phase. The XENON100 experiment has released data taken in the years 2011–2012 for an exposure of 224.6 days, using a fiducial volume of just 34 kg of Xenon target mass [27]. See related discussions in literature about the detector response of such devices, in particular, to low energy recoils [11, 12, 28]. The technical performance of the apparata, confirmed also by similar experiments, has shown e.g. that: i) the detectors are affected by large non-uniformity; some kind of corrections may be estimated and applied, but significant systematics has to be accounted for; ii) the response of these detectors is not linear, i.e. the number of photoelectrons/keV depends on the energy scale and depends also on the applied electric field; iii) the physical energy threshold is not suitably proved iv) the use of energy calibration lines from Xe activated by neutrons cannot be applied as routine and the studies on a possible calibration with internal sources in the same running conditions have not been realized so far; v) despite of the small light response (2.28 photoelectron/keVee), an energy threshold at 1.3 keVee is claimed; vi) the energy resolution is poor; vii) in the scale-up of the detectors the performances deteriorate; viii) the behaviour of the light yield for recoils at low energy is uncertain in every case. LUX – a dual phase TPC filled with Xenon – reports the first results corresponding to an exposure of 85.3 days, using a fiducial volume of 118 kg [29]; the last data release refers to about 1.4×10^4 kg $\times$ day [30]. On the other hand, the first result of DARKSIDE has been obtained with the TPC filled with atmospheric Argon for an exposure of 1422 kg $\times$ day [31]. Similar considerations, as above, hold about the robustness of these results. Moreover, such considerations become still more restrictive when considering the future plans of larger set-ups [11, 12].

A positive hint for a signal of light DM candidates inducing just nuclear elastic scatterings has been also reported by the CoGeNT experiment [32, 33]. The set-up is composed by a 440 g, p-type point contact (PPC) Ge diode, with a very low energy threshold at 0.4 keVee. It is located in the Soudan Underground Laboratory. In the data analysis no discrimination between electromagnetic radiation and nuclear recoils is applied; only noise events are rejected. The experiment observes more events than they expect from estimate of the background in the energy range 0.4-3.2 keVee. The energy spectrum of these events is compatible – under certain assumptions – with a signal produced by the interaction of a DM particle with a mass around 10 GeV. In addition, considering an exposure of 146 kg $\times$ days CoGeNT experiment also reports an evidence at about 2.2σ C.L. of an annual modulation of the counting rate in (0.5-2) keVee with phase and period compatible – although the small confidence level – with a DM signal [33]. This result is compatible with interpretations of the DM model-independent annual modulation result already reported by DAMA in terms of this kind of DM candidate and with the possible hint reported above. A new data release is planned in the incoming months, and CoGeNT is upgrading towards C-4 with the aim to improve by a factor four the total mass, to decrease the total background and to reduce substantially the energy threshold; Soudan is still the laboratory. Other activities exploiting Ge detectors are Texono and CDEX at CJPL, the Chinese underground laboratory.

Finally, activities using inorganic scintillators are in progress and at the R&D stage; in particular here we remind the efforts of the long-standing ANAIS project whose goal is to run about 100 kg of NaI(Tl) at Canfranc laboratory in Spain [34].

In conclusion, suitable experiments offering a model independent signature for the presence of DM particles in the galactic halo are mandatory.

IOP Conf. Series: Journal of Physics: Conf. Series **866** (2017) 012016 doi:10.1088/1742-6596/866/1/012016

3. DM model independent signature

To obtain a reliable signature for the presence of DM particles in the galactic halo, it is necessary to exploit a suitable model independent signature. With the present technology, one feasible and able to test a large range of cross sections and of DM particle halo densities, is the so-called DM annual modulation signature [35]. The annual modulation of the signal rate originates from the Earth revolution around the Sun. In fact, as a consequence of its annual revolution around the Sun, which is moving in the Galaxy traveling with respect to the Local Standard of Rest towards the star Vega near the constellation of Hercules, the Earth should be crossed by a larger flux of DM particles around ∼2 June (when the Earth orbital velocity is summed to the one of the solar system with respect to the Galaxy) and by a smaller one around ∼2 December (when the two velocities are subtracted). Thus, this signature has a different origin and peculiarities than the seasons on the Earth and than effects correlated with seasons (consider the expected value of the phase as well as the other requirements listed below). This DM annual modulation signature is very distinctive since the effect induced by DM particles must simultaneously satisfy all the following requirements: (1) the rate must contain a component modulated according to a cosine function; (2) with one year period; (3) with a phase that peaks roughly around ∼ 2nd June; (4) this modulation must be present only in a well-defined low energy range, where DM particles can induce signals; (5) it must be present only in those events where just a single detector, among all the available ones in the used set-up, actually "fires" (*single-hit* events), since the probability that DM particles experience multiple interactions is negligible; (6) the modulation amplitude in the region of maximal sensitivity has to be 7% in case of usually adopted halo distributions, but it may be significantly larger in case of some particular scenarios such as e.g. those in Ref. [6, 36]. This signature is model independent and might be mimicked only by systematic effects or side reactions able to simultaneously satisfy all the requirements given above; no one is available. At present status of technology it is the only DM model independent signature available in direct DM investigation that can be effectively exploited.

4. DAMA DM annual modulation results with highly radiopure NaI(Tl)

The DM annual modulation signature has been exploited with large exposure – using highly radiopure NaI(Tl) as target material – by the former DAMA/NaI ($\simeq$ 100 kg sensitive mass) experiment [1, 8, 9, 10, 37, 38, 39, 40, 41, 42], and by the currently running DAMA/LIBRA ($\simeq$ 250 kg sensitive mass) [43, 44, 45, 46, 47, 48, 49, 50, 51, 52, 53, 54, 55, 56, 57], within the DAMA project. The DAMA project is dedicated to the development and use of low background scintillators for underground physics.

In particular, the experimental observable in DAMA experiments is the modulated component of the signal in NaI(Tl) target and not the constant part of it, as done in the other approaches aforementioned.

The full description of the DAMA/LIBRA set-up and performances during the phase1 and phase2 (presently running) and other related arguments have been discussed in details in Refs. [43, 44, 45, 46, 48, 49, 55, 50] and references therein. Here we just remind that the sensitive part of this set-up is made of 25 highly radiopure NaI(Tl) crystal scintillators (5-rows by 5-columns matrix) having 9.70 kg mass each one. In each detector two 10 cm long UV light guides (made of Suprasil B quartz) act also as optical windows on the two end faces of the crystal, and are coupled to two low background photomultipliers (PMTs) working in coincidence at single photoelectron level. The low background 9265-B53/FL and 9302-A/FL PMTs, developed by EMI-Electron Tubes with dedicated R&Ds, were used in the phase1; for details see Ref. [43, 37, 39] and references therein. The detectors are housed in a sealed low-radioactive copper box installed in the center of a low-radioactive Cu/Pb/Cd-foils/polyethylene/paraffin shield; moreover, about 1 m concrete (made from the Gran Sasso rock material) almost fully surrounds (mostly outside the barrack) this passive shield, acting as a further neutron moderator. A threefold-levels sealing

IOP Conf. Series: Journal of Physics: Conf. Series **866** (2017) 012016 doi:10.1088/1742-6596/866/1/012016

system prevents the detectors to be in contact with the environmental air of the underground laboratory [43]. The light response of the detectors during phase1 typically ranges from 5.5 to 7.5 photoelectrons/keV, depending on the detector. The hardware threshold of each PMT is at single photoelectron, while a software energy threshold of 2 keV electron equivalent (hereafter keV) is used [43, 37]. Energy calibration with X-rays/γ sources are regularly carried out in the same running condition down to few keV [43]; in particular, double coincidences due to internal X-rays from ^{40}K (which is at ppt levels in the crystals) provide (when summing the data over long periods) a calibration point at 3.2 keV close to the software energy threshold (for details see Ref. [43]). The radiopurity, the procedures and details are discussed in Ref. [43, 44, 45, 46, 50] and references therein.

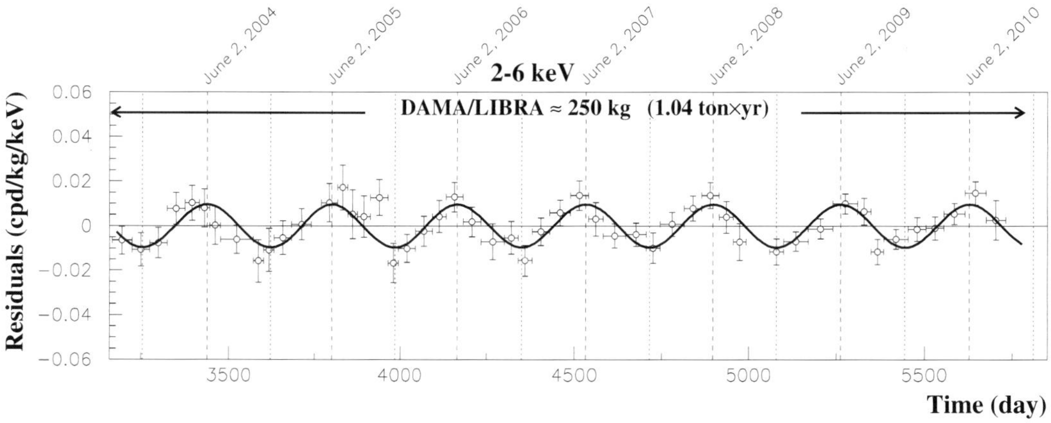

Figure 1. Experimental residual rate of the *single-hit* scintillation events measured by DAMA/LIBRA–phase1 in the (2–6) keV energy interval as a function of the time. The superimposed curve is the cosinusoidal function behaviour $A\cos\omega(t-t_0)$ with a period $T = \frac{2\pi}{\omega} = 1$ yr, a phase $t_0 = 152.5$ day (June 2nd) and modulation amplitude, A, equal to the central values obtained by best fit on the data points of the entire DAMA/LIBRA–phase1. The dashed vertical lines correspond to the maximum expected for the DM signal (June 2nd), while the dotted vertical lines correspond to the minimum.

The data of DAMA/LIBRA–phase1 correspond to 1.04 ton $\times$ yr collected in 7 annual cycles; when including also the data of the DAMA/NaI experiment the total exposure is 1.33 ton $\times$ yr collected in 14 annual cycles. In order to investigate the presence of an annual modulation with proper features in the data, many analyses have been carried out. All these analyses point out the presence of an annual modulation satisfying all the requirements of the signature [44, 45, 46, 50]. In Fig. 1, as example, it is plotted the time behaviour of the experimental residual rate of the *single-hit* scintillation events for DAMA/LIBRA–phase1 in the (2–6) keV energy interval. When fitting the *single-hit* residual rate of DAMA/LIBRA–phase1 together with the DAMA/NaI ones, with the function: $A\cos\omega(t-t_0)$, considering a period $T = \frac{2\pi}{\omega} = 1$ yr and a phase $t_0 = 152.5$ day (June 2nd) as expected by the DM annual modulation signature, the following modulation amplitude is obtained: $A = (0.0110 \pm 0.0012)$ cpd/kg/keV, corresponding to 9.2 σ C.L..

When the period, and the phase are kept free in the fitting procedure, the modulation amplitude is (0.0112 ± 0.0012) cpd/kg/keV (9.3 σ C.L.), the period $T = (0.998 \pm 0.002)$ year and the phase $t_0 = (144 \pm 7)$ day, values well in agreement with expectations for a DM annual modulation signal. In particular, the phase is consistent with about June 2nd and is fully consistent with the value independently determined by Maximum Likelihood analysis [46].

For completeness, we recall that a slight energy dependence of the phase could be expected in case of possible contributions of non-thermalized DM components to the galactic halo, such as e.g. the SagDEG stream [41, 58, 59] and the caustics [60]. For more details see Ref. [46].

The modulation amplitudes singularly calculated for each annual cycle of DAMA/NaI and DAMA/LIBRA–phase1 are compatible among them and are normally fluctuating around their best fit values [44, 45, 46]. In particular, for the (2–6) keV energy interval the χ^2 is 10.8 over 13 *d.o.f.* corresponding to an upper tail probability of 63%, while the *run test* yields a lower tail probabilities of 23%. This analysis confirms that the data collected in all the annual cycles with DAMA/NaI and DAMA/LIBRA–phase1 are statistically compatible and can be considered together.

The DAMA/LIBRA–phase1 *single-hit* residuals of Fig. 1 and those of DAMA/NaI have also been investigated by a Fourier analysis. The data analysis procedure has been described in details in Ref. [50]. A clear peak corresponding to a period of 1 year (see Fig. 2 – *left*) is evident for the (2–6) keV energy interval; the same analysis in the (6–14) keV energy region shows instead only aliasing peaks. Neither other structure at different frequencies has been observed (see also Ref. [50]).

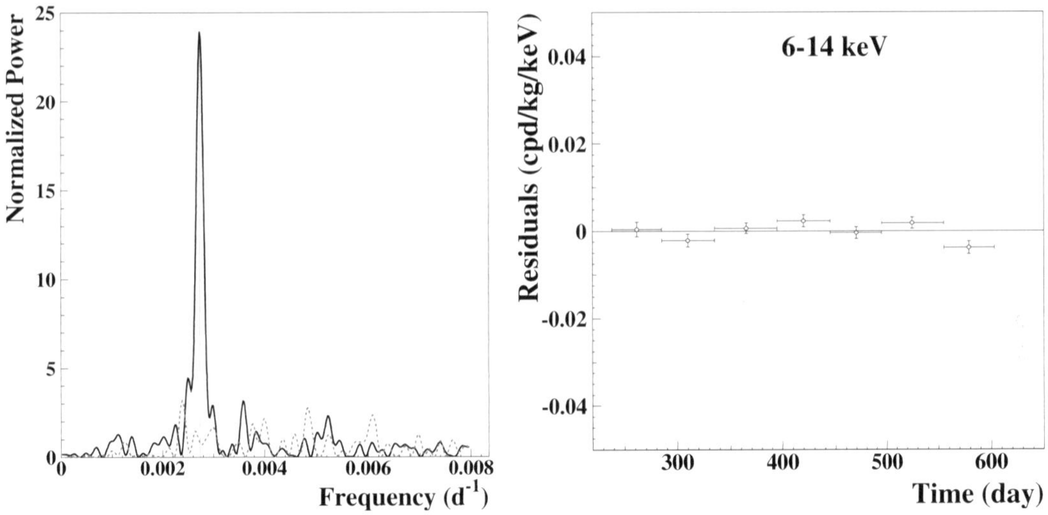

Figure 2. *Left:* Power spectrum of the measured *single-hit* residuals in the (2–6) keV (solid lines) and (6–14) keV (dotted lines) energy intervals calculated according to Ref. [50], including also the treatment of the experimental errors and of the time binning. The data refer to DAMA/NaI and DAMA/LIBRA–phase1. As it can be seen, the principal mode present in the (2–6) keV energy interval corresponds to a frequency of 2.737×10^{-3} d^{-1} (vertical lines), corresponding to a period of $\simeq$ 1 year. A similar peak is not present in the (6–14) keV energy interval. *Right:* Experimental *single-hit* residuals in the (6–14) keV energy region for the entire DAMA/LIBRA–phase1 data as if they were collected in a single annual cycle (i.e. binning in the variable time from the Jan 1st of each annual cycle). The data points present the experimental errors as vertical bars and the associated time bin width as horizontal bars. The initial time of the figures is taken at August 7th. The amplitude is well compatible with zero: A=(0.00032 $\pm$ 0.00076) cpd/kg/keV.

Absence of any other significant background modulation in the energy spectrum has been verified in energy regions not of interest for DM; e.g. the measured rate integrated above 90 keV, R_{90}, as a function of the time has been analysed [46]. Similar result is obtained in other energy intervals; for example Fig. 2 – *right* shows the *single-hit* residuals in the (6–14) keV energy region for the entire DAMA/LIBRA–phase1 data as if they were collected in a single annual cycle (i.e. binning in the variable time from the Jan 1st of each annual cycle). It is worth noting that the obtained results account of whatever kind of background and, in addition, no background process able to mimic the DM annual modulation signature (that is able to simultaneously satisfy all the peculiarities of the signature and to account for the measured modulation amplitude) is available (see also discussions e.g. in Ref. [43, 44, 45, 46, 49, 50, 54, 61, 62, 63, 64, 65, 66, 67]).

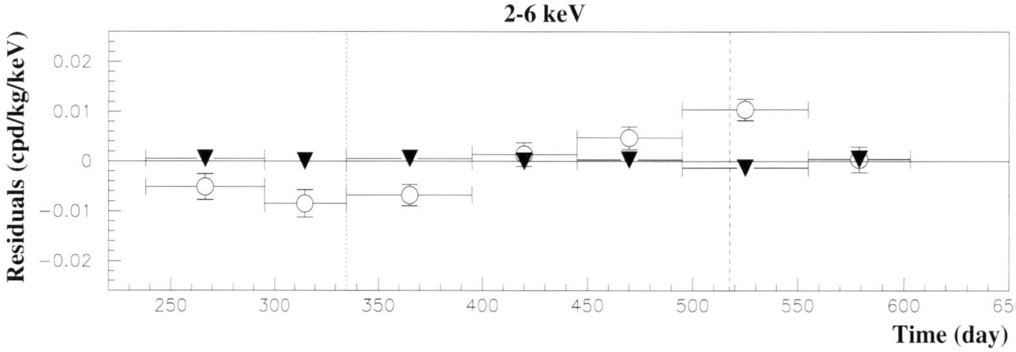

Figure 3. Experimental residual rates of DAMA/LIBRA–phase1 *single-hit* events (open circles), class of events to which DM events belong, and for *multiple-hit* events (filled triangles), class of events to which DM events do not belong. They have been obtained by considering for each class of events the data as collected in a single annual cycle and by using in both cases the same identical hardware and the same identical software procedures. The initial time of the figure is taken on August 7th. The experimental points present the errors as vertical bars and the associated time bin width as horizontal bars. Analogous results were obtained for the DAMA/NaI data [40].

A further relevant investigation in the DAMA/LIBRA–phase1 data has been performed by applying the same hardware and software procedures, used to acquire and to analyse the *single-hit* residual rate, to the *multiple-hit* one. In fact, since the probability that a DM particle interacts in more than one detector is negligible, a DM signal can be present just in the *single-hit* residual rate. Thus, the comparison of the results of the *single-hit* events with those of the *multiple-hit* ones corresponds practically to compare between them the cases of DM particles beam-on and beam-off. This procedure also allows an additional test of the background behaviour in the same energy interval where the positive effect is observed. In particular, in Fig. 3 the residual rates of the *single-hit* events measured over the DAMA/LIBRA–phase1 annual cycles are reported, as collected in a single cycle, together with the residual rates of the *multiple-hit* events, in the (2–6) keV energy interval. While, as already observed, a clear modulation, satisfying all the peculiarities of the DM annual modulation signature, is present in the *single-hit* events, the fitted modulation amplitude for the *multiple-hit* residual rate is well compatible with zero: $-(0.0005 \pm 0.0004)$ cpd/kg/keV in the energy region (2–6) keV. Thus, again evidence of annual modulation with the features required by the DM annual modulation signature is present in the *single-hit* residuals (events class to which the DM particle induced events belong), while it is absent in the *multiple-hit* residual rate (event class to which only background events belong). Similar results were also obtained for the last two annual cycles of the DAMA/NaI experiment [40]. Since the same identical hardware and the same identical software procedures have been used to analyse the two classes of events, the obtained result offers an additional strong support for the presence of a DM particle component in the galactic halo.

The annual modulation present at low energy can also be pointed out by depicting – as a function of the energy – the modulation amplitude, $S_{m,k}$, obtained by maximum likelihood method considering T =1 yr and t_0 = 152.5 day. For such purpose the likelihood function of the *single-hit* experimental data in the $k-$th energy bin is defined as: $\mathbf{L_k} = \mathbf{\Pi}_{ij} e^{-\mu_{ijk}} \frac{\mu_{ijk}^{N_{ijk}}}{N_{ijk}!}$, where N_{ijk} is the number of events collected in the i-th time interval (hereafter 1 day), by the j-th detector and in the k-th energy bin. N_{ijk} follows a Poisson's distribution with expectation value $\mu_{ijk} = [b_{jk} + S_{ik}] M_j \Delta t_i \Delta E \epsilon_{jk}$. The b_{jk} are the background contributions, M_j is the mass of the $j-$th detector, Δt_i is the detector running time during the i-th time interval, ΔE is the chosen energy bin, ϵ_{jk} is the overall efficiency. Moreover, the signal can be written as $S_{ik} = S_{0,k} + S_{m,k} \cdot \cos\omega(t_i - t_0)$, where $S_{0,k}$ is the constant part of the signal and $S_{m,k}$ is the modulation amplitude. The usual procedure is to minimize the function $y_k = -2ln(\mathbf{L_k}) - const$ for each energy bin; the free parameters of the fit are the $(b_{jk} + S_{0,k})$ contributions and the $S_{m,k}$ parameter. Hereafter, the index k is omitted for simplicity.

In Fig. 4 the obtained S_m are shown in each considered energy bin (there $\Delta E = 0.5$ keV) when the data of DAMA/NaI and DAMA/LIBRA–phase1 are considered. It can be inferred that positive signal is present in the (2–6) keV energy interval, while S_m values compatible with zero are present just above. In fact, the S_m values in the (6–20) keV energy interval have random fluctuations around zero with χ^2 equal to 35.8 for 28 degrees of freedom (upper tail probability of 15%). All this confirms the previous analyses.

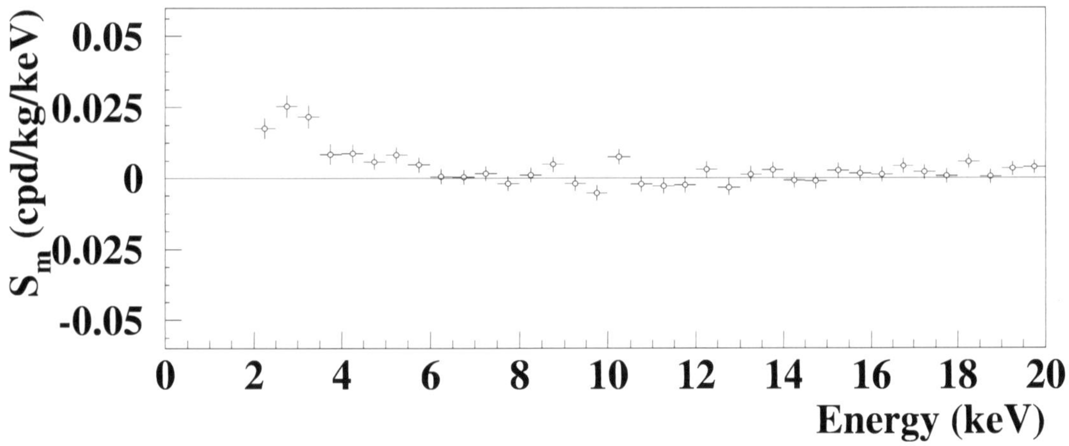

Figure 4. Energy distribution of the S_m variable for the total cumulative exposure 1.33 ton×yr. The energy bin is 0.5 keV. A clear modulation is present in the lowest energy region, while S_m values compatible with zero are present just above. In fact, the S_m values in the (6–20) keV energy interval have random fluctuations around zero with χ^2 equal to 35.8 for 28 degrees of freedom (upper tail probability of 15%).

As described in Ref. [44, 45, 46, 50], the observed annual modulation effect is well distributed in all the 25 detectors at 95% C.L.

Among further additional tests, the analysis of the modulation amplitudes as a function of the energy separately for the nine inner detectors and the remaining external ones has been carried out for the entire DAMA/LIBRA–phase1. The obtained values are fully in agreement; in fact, the hypothesis that the two sets of modulation amplitudes as a function of the energy belong to same distribution has been verified by χ^2 test, obtaining: $\chi^2/d.o.f. = 3.9/4$ and 8.9/8 for the energy intervals (2–4) and (2–6) keV, respectively (ΔE = 0.5 keV). This shows that the effect is also well shared between inner and outer detectors.

Let us, finally, release the assumption of a phase $t_0 = 152.5$ day in the procedure to evaluate the modulation amplitudes. In this case the signal can be written as:

$$\begin{aligned} S_{ik} &= S_{0,k} + S_{m,k} \cos\omega(t_i - t_0) + Z_{m,k} \sin\omega(t_i - t_0) \\ &= S_{0,k} + Y_{m,k} \cos\omega(t_i - t^*). \end{aligned} \quad (1)$$

For signals induced by DM particles one should expect: i) $Z_{m,k} \sim 0$ (because of the orthogonality between the cosine and the sine functions); ii) $S_{m,k} \simeq Y_{m,k}$; iii) $t^* \simeq t_0 = 152.5$ day. In fact, these conditions hold for most of the dark halo models; however, as mentioned above, slight differences can be expected in case of possible contributions from non-thermalized DM components, such as e.g. the SagDEG stream [41, 58, 59] and the caustics [60].

Table 1. Best fit values for the (2–6) and (6–14) keV energy intervals (1σ errors) for S_m versus Z_m and Y_m versus t^*, considering the cumulative exposure of DAMA/NaI and DAMA/LIBRA–phase1. See also Fig. 5.

E (keV)	S_m (cpd/kg/keV)	Z_m (cpd/kg/keV)	Y_m (cpd/kg/keV)	t^* (day)
2–6	(0.0106 ± 0.0012)	-(0.0006 ± 0.0012)	(0.0107 ± 0.0012)	(149.5 ± 7.0)
6–14	(0.0001 ± 0.0007)	(0.0000 ± 0.0005)	(0.0001 ± 0.0008)	undefined

Considering cumulatively the data of DAMA/NaI and DAMA/LIBRA–phase1 the obtained 2σ contours in the plane (S_m, Z_m) for the (2–6) keV and (6–14) keV energy intervals are shown in Fig. 5–*left* while in Fig. 5–*right* the obtained 2σ contours in the plane (Y_m, t^*) are depicted. The best fit values for the (2–6) and (6–14) keV energy intervals (1σ errors) for S_m versus Z_m and Y_m versus t^* are reported in Table 1.

Finally, setting S_m in eq. (1) to zero, the Z_m values as function of the energy have also been determined by using the same procedure. The values of Z_m are well compatible with zero, as expected [44, 45, 46].

No modulation has been found in any possible source of systematics or side reactions; thus, cautious upper limits on possible contributions to the DAMA/LIBRA–phase1 measured modulation amplitude have been obtained (see Refs. [44, 45, 46, 38, 39, 40, 49, 55]). It is worth noting that they do not quantitatively account for the measured modulation amplitudes, and also are not able to simultaneously satisfy all the many requirements of the signature. Similar analyses have also been performed for the DAMA/NaI data [39, 40].

Sometimes naive statements were put forward as the fact that in nature several phenomena may show some kind of periodicity. It is worth noting that the point is whether they might mimic the annual modulation signature in DAMA/NaI and in DAMA/LIBRA, i.e. whether they might be not only quantitatively able to account for the observed modulation amplitude but also able to contemporaneously satisfy all the requirements of the DM annual modulation signature. The same is for side reactions too. This has already been deeply investigated and discussed in DAMA literature.

In particular, in Refs. [49, 55] a quantitative evaluation why the neutrons, the muons and the solar neutrinos cannot give any significant contribution to the DAMA annual modulation results and cannot mimic this signature is outlined. Table 2 summarizes the safety upper limits on the contributions to the observed modulation amplitude due to the total neutron flux at LNGS, either from (α, n) reactions, from fissions and from muons' and solar-neutrinos' interactions in the rocks and in the lead around the experimental set-up; the direct contributions of muons and solar neutrinos are reported there too.

In any case no systematic effects or side reactions able to account for the whole observed modulation amplitude and to simultaneously satisfy all the requirements of the exploited DM

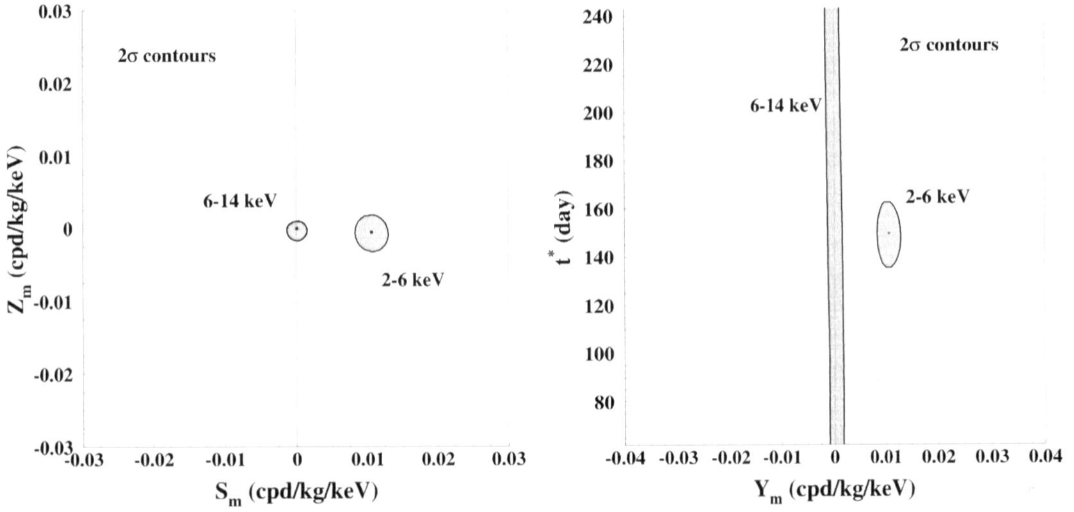

Figure 5. 2σ contours in the plane (S_m, Z_m) (*left*) and in the plane (Y_m, t^*) (*right*) for the (2–6) keV and (6–14) keV energy intervals. The contours have been obtained by the maximum likelihood method, considering the cumulative exposure of DAMA/NaI and DAMA/LIBRA–phase1. A modulation amplitude is present in the lower energy intervals and the phase agrees with that expected for DM induced signals. See text.

Table 2. Summary of the contributions to the total neutron flux at LNGS; the value, $\Phi^{(n)}_{0,k}$, the relative modulation amplitude, η_k, and the phase, t_k, of each component is reported. It is also reported the counting rate, $R_{0,k}$, in DAMA/LIBRA for *single-hit* events, in the $(2-6)$ keV energy region induced by neutrons, muons and solar neutrinos, detailed for each component. The modulation amplitudes, A_k, are reported as well, while the last column shows the relative contribution to the annual modulation amplitude observed by DAMA/LIBRA, $S_m^{exp} \simeq 0.0112$ cpd/kg/keV [46]. For details see Ref. [55] and references therein.

Source		$\Phi^{(n)}_{0,k}$ (neutrons cm^{-2} s^{-1})	η_k	t_k	$R_{0,k}$ (cpd/kg/keV)	$A_k = R_{0,k}\eta_k$ (cpd/kg/keV)	A_k/S_m^{exp}
SLOW neutrons	thermal n ($10^{-2} - 10^{-1}$ eV)	1.08×10^{-6}	$\simeq 0$ however $\ll 0.1$	–	$< 8 \times 10^{-6}$	$\ll 8 \times 10^{-7}$	$\ll 7 \times 10^{-5}$
	epithermal n (eV-keV)	2×10^{-6}	$\simeq 0$ however $\ll 0.1$	–	$< 3 \times 10^{-3}$	$\ll 3 \times 10^{-4}$	$\ll 0.03$
FAST neutrons	fission, $(\alpha, n) \to$ n (1-10 MeV)	$\simeq 0.9 \times 10^{-7}$	$\simeq 0$ however $\ll 0.1$		$< 6 \times 10^{-4}$	$\ll 6 \times 10^{-5}$	$\ll 5 \times 10^{-3}$
	$\mu \to$ n from rock (> 10 MeV)	$\simeq 3 \times 10^{-9}$	0.0129	end of June	$\ll 7 \times 10^{-4}$	$\ll 9 \times 10^{-6}$	$\ll 8 \times 10^{-4}$
	$\mu \to$ n from Pb shield (> 10 MeV)	$\simeq 6 \times 10^{-9}$	0.0129	end of June	$\ll 1.4 \times 10^{-3}$	$\ll 2 \times 10^{-5}$	$\ll 1.6 \times 10^{-3}$
	$\nu \to$ n (few MeV)	$\simeq 3 \times 10^{-10}$	0.03342*	Jan. 4th*	$\ll 7 \times 10^{-5}$	$\ll 2 \times 10^{-6}$	$\ll 2 \times 10^{-4}$
	direct μ	$\Phi_0^{(\mu)} \simeq 20\ \mu$ m^{-2}d^{-1}	0.0129	end of June	$\simeq 10^{-7}$	$\simeq 10^{-9}$	$\simeq 10^{-7}$
	direct ν	$\Phi_0^{(\nu)} \simeq 6 \times 10^{10}\ \nu$ cm^{-2}s^{-1}	0.03342*	Jan. 4th*	$\simeq 10^{-5}$	3×10^{-7}	3×10^{-5}

* The annual modulation of solar neutrino is due to the different Sun-Earth distance along the year; so the relative modulation amplitude is twice the eccentricity of the Earth orbit and the phase is given by the perihelion.

signature have been found. A detailed discussion about all the related arguments can be found in Refs. [43, 44, 45, 46, 49, 50, 54, 38, 39, 40, 55].

Analyses on the presence of possible diurnal effects in the DAMA/LIBRA–phase1 data [54] and on the so called "Earth Shadow Effect" [56] will be summarized in the following. Finally, the annual modulation result has been also interpreted in terms of Asymmetric Mirror DM [57], as briefly shown later.

5. Implications and comparisons

The long-standing annual-modulation evidence measured in DAMA experiments is model-independent, i.e. in particular independent on theoretical interpretations of the identity of DM and specifics of its interactions. It can be related to a variety of interaction mechanisms of DM particles with the detector materials and is compatible with a wide set of scenarios regarding the nature of the DM candidate and related astrophysical, nuclear and particle Physics. For example, some of the scenarios available in literature and the different parameters are discussed in Refs. [39, 40, 10, 41, 1, 42, 9, 8, 44, 50] and references therein, and recently e.g. in Refs. [68, 57] . Further large literature is available on the topics (see for example in Ref. [50]) and many possibilities are open.

It is worth noting that no other experiment exists, whose result can be directly compared in a model-independent way with those by DAMA/NaI and DAMA/LIBRA. Some activities claim model-dependent exclusion under many largely arbitrary assumptions (see for example discussions in Ref. [39, 44, 40, 11, 12, 28]). Moreover, often some critical points exist in their experimental aspects, as mentioned above, and the existing experimental and theoretical uncertainties are generally not considered in their presented single model dependent result; moreover, implications of the DAMA results are often presented in incorrect/partial/unupdated way. Both the accounting of the existing uncertainties and the existence of alternative scenarios (see literature) allow one to note that model dependent results by indirect and direct experiments actually are not in conflict with the DAMA model independent result.

5.1. The case of Asymmetric Mirror DM

The model independent annual modulation effect observed by the DAMA experiments has been investigated in terms of many DM candidates. Here we just recall the case of a mirror-type dark matter candidates in some scenarios [57].

In the framework of asymmetric mirror matter, the DM originates from hidden (or shadow) gauge sectors which have particles and interaction content similar to that of ordinary particles. It is assumed that the mirror parity is spontaneously broken and the electroweak symmetry breaking scale v' in the mirror sector is much larger than that in the Standard Model, $v = 174$ GeV. In this case, the mirror world becomes a heavier and deformed copy of our world, with mirror particle masses scaled in different ways with respect to the masses of the ordinary particles. Then, in this scenario dark matter would exist in the form of mirror hydrogen composed of mirror proton and electron, with mass of about 5 GeV which is a rather interesting mass range for dark matter particles.

The data analysis in the Mirror DM model framework allows the determination of the $\sqrt{f}\epsilon$ parameter (where f is the fraction of DM in the Galaxy in form of mirror atoms and ϵ is the coupling constant). In the analysis several uncertainties on the astrophysical, particle physics and nuclear physics models have been taken into account in the calculation. The obtained values of the $\sqrt{f}\epsilon$ parameter in the case of mirror hydrogen atom ranges between 7.7×10^{-10} to 1.1×10^{-7}; they are well compatible with cosmological bounds [57].

In addition, releasing the assumption $M_{A'} \simeq 5m_p$, the allowed regions for the $\sqrt{f}\epsilon$ parameter as function of $M_{A'}$, mirror hydrogen mass, obtained by marginalizing all the models for each

IOP Conf. Series: Journal of Physics: Conf. Series **866** (2017) 012016 doi:10.1088/1742-6596/866/1/012016

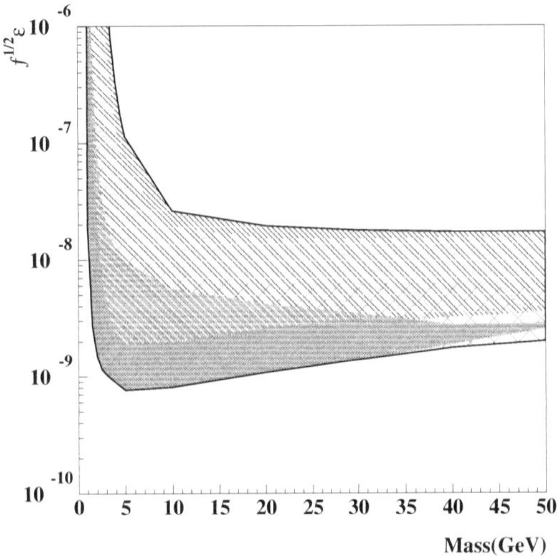

Figure 6. Allowed regions for the $\sqrt{f}\epsilon$ parameter as function of mirror hydrogen mass, obtained by marginalizing all the models for each considered scenario. The allowed intervals identify the $\sqrt{f}\epsilon$ values corresponding to C.L. larger than 5σ from the *null hypothesis*, that is $\sqrt{f}\epsilon = 0$. The allowed regions corresponding to five different scenarios are depicted in different hatching; the black line is the overall boundary; for details see Ref. [57].

considered scenario, are shown in Fig. 6. Five scenarios are reported with different hatching of the allowed regions.

6. Other signatures?

In the framework of direct DM investigations crystal scintillators represent a reliable technology to investigate model independent signatures. An apparatus made by an array of such detectors, in addition to the already mentioned advantages, allows a long term stability and effective routine calibrations down to keV energy region in the same conditions as production runs. Moreover due to the wide choice of nuclei and isotopes, they are sensitive to many candidates, interaction types and astrophysical, nuclear and particle physics scenarios. In particular in order to disentangle in the corollary investigation on the candidate particle(s) at least some of the many possible astrophysical, nuclear and particle Physics scenarios and related experimental and theoretical uncertainties, the decreasing of the software energy threshold and higher exposures are necessary.

Finally, the only effective method for such studies is the investigation of a model independent signature and of second order effects.

Many topics can be investigated:

- the peculiarities of the annual modulation phase;
- the peculiarities of the DM interaction mechanisms;
- the velocity and spatial distribution of the DM particles in the galactic halo;
- the effects induced on the DM particles distribution in the galactic halo by contributions from satellite galaxies tidal streams;
- the effects induced on the DM particles distribution in the galactic halo by the possible existence of caustics;
- the detection of possible solar wakes or gravitational focusing effect of the Sun on the DM particle;

- the investigation of possible diurnal effects and so on.

Some of these studies already started with DAMA/LIBRA–phase1([54, 50, 69] and references therein), but a further important step towards such investigations will be represented by the larger exposure collected by DAMA/LIBRA–phase2, with lower energy threshold.

6.1. Diurnal modulation

The results obtained by investigating the presence of possible diurnal variation in the low-energy *single-hit* scintillation events collected by DAMA/LIBRA–phase1 (1.04 ton × year exposure) have been analysed in terms of a DM second order model-independent effect due to the Earth diurnal rotation around its axis [54]. In particular, the data were analysed using the sidereal time referred to Greenwich, often called GMST.

This daily modulation of the rate on the sidereal time, expected when taking into account the contribution of the Earth rotation velocity, has several requirements as the DM annual modulation effect does. The interest in this signature is that the ratio R_{dy} of this diurnal modulation amplitude over the annual modulation amplitude is a model independent constant at given latitude; considering the LNGS latitude one has:

$$R_{dy} = \frac{S_d}{S_m} \simeq 0.016 \qquad (2)$$

Taking into account R_{dy} and the DM annual modulation effect pointed out by DAMA/LIBRA–phase1 for *single-hit* events in the low energy region, it is possible to derive the diurnal modulation amplitude expected for the same data. In particular, when considering the (2–6) keV energy interval, the observed annual modulation amplitude in DAMA/LIBRA–phase1 is: (0.0097 ± 0.0013) cpd/kg/keV [46] and the expected value of the diurnal modulation amplitude is $\simeq 1.5 \times 10^{-4}$ cpd/kg/keV.

Fig. 7 shows the time and energy behaviour of the experimental residual rates of *single-hit* events both as a function of solar (*left*) and of sidereal (*right*) time, in the (2–6) keV and (6–14) keV intervals. The used time bin is 1 (either solar or sidereal) hour.

The null hypothesis (absence of residual rate diurnal variation) has been tested by a χ^2 test, obtaining the results given in Table 3; there the upper tail probabilities (P-values), calculated by the standard χ^2 distribution, are also reported. Thus, no diurnal variation with a significance of 95% C.L. is found at the reached level of sensitivity.

Table 3. Test of absence of diurnal effect in the DAMA/LIBRA–phase1 data. The P-values, calculated by the standard χ^2 distribution, are also shown. As can be seen, the χ^2 test supports the hypothesis that the diurnal residual rates in DAMA/LIBRA–phase1 are simply fluctuating around zero.

Energy	Solar Time	Sidereal Time
2–6 keV	χ^2/d.o.f. = 25.8/24 → P = 36%	χ^2/d.o.f. = 21.2/24 → P = 63%
6–14 keV	χ^2/d.o.f. = 25.5/24 → P = 38%	χ^2/d.o.f. = 35.9/24 → P = 6%

In addition to the χ^2 test, another independent statistical test has been applied: the run test [54] ; it verifies the hypothesis that the positive and negative data points are randomly distributed. The lower tail probabilities are equal to: 7% and 26% in the (2–6) and (6-14) keV energy region, respectively, for the solar case and 78% and 16% in the (2–6) and (6-14) keV energy region, respectively, for the sidereal case. Thus, in conclusion the presence of any significant diurnal variation and of time structures can be excluded at the reached level of sensitivity (see Fig. 7).

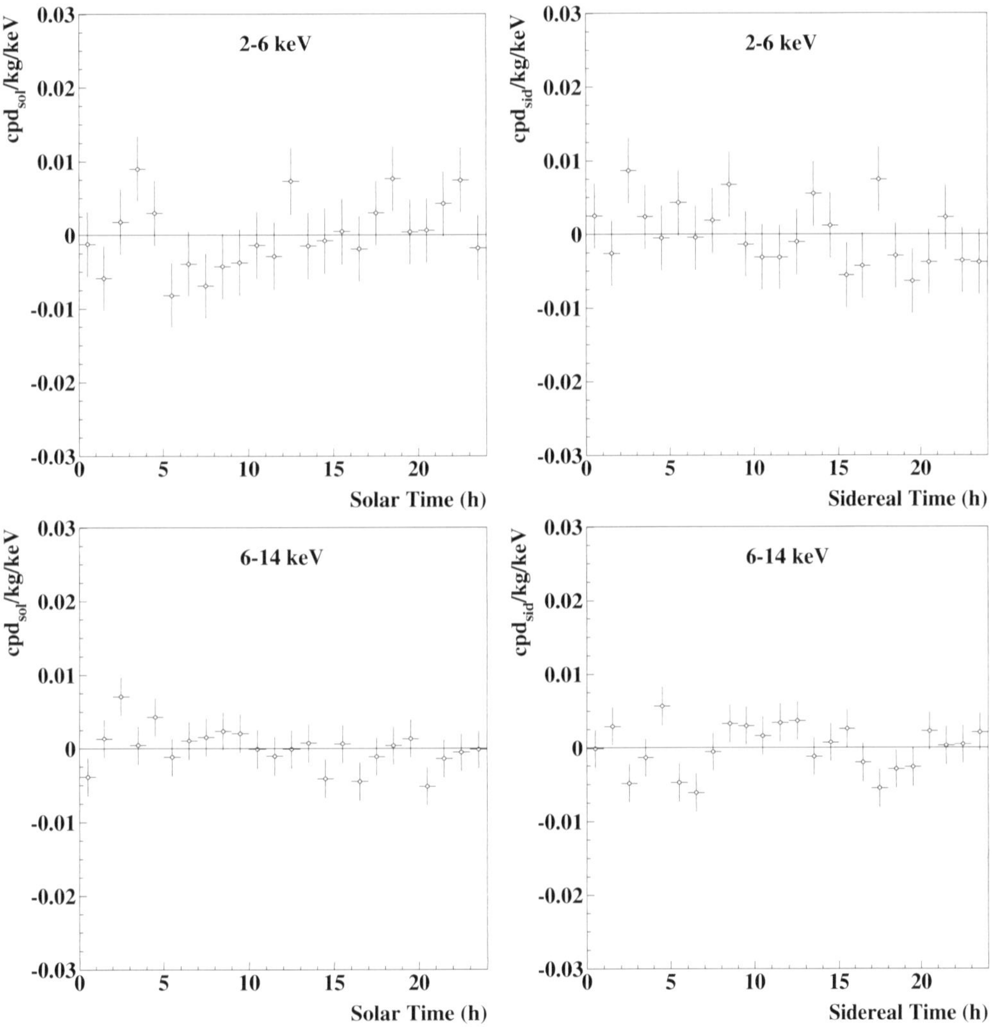

Figure 7. Experimental model-independent diurnal residual rate of the *single-hit* scintillation events, measured by DAMA/LIBRA–phase1 in the (2–6) and (6–14) keV energy intervals as a function of the hour of the solar ($left$) and sidereal ($right$) day. The experimental points present the errors as vertical bars and the associated time bin width (1 hour) as horizontal bars. The cumulative exposure is 1.04 ton × yr. See Ref. [54] for details.

In order to compare the experimental data with the DM diurnal effect due to the Earth rotation around its axis, the sidereal diurnal modulation amplitude of the (2–6) keV energy interval is taken into account: $A_d^{exp} = -(1.0 \pm 1.3) \times 10^{-3}$ cpd/kg/keV. Following the Feldman-Cousins procedure an upper limit can be obtained for the measured diurnal modulation amplitude: $A_d^{exp} < 1.2 \times 10^{-3}$ cpd/kg/keV (90% C.L.); thus, the effect of DM diurnal modulation (expected amplitude $\simeq 1.5\times\ 10^{-4}$ cpd/kg/keV) is out the present sensitivity [54].

In conclusion, at that level of sensitivity of DAMA/LIBRA–phase1 the presence of a significant diurnal variation and of diurnal time structures in the data can be excluded for both the cases of solar and sidereal time. In particular, the sidereal diurnal modulation amplitude expected – because of the Earth diurnal motion – on the basis of the DAMA DM annual modulation results cannot be investigated at the present sensitivity; DAMA/LIBRA–phase2, presently running, with a lower software energy threshold [48] can also offer the possibility to increase sensitivity to such an effect.

6.2. Daily effect on the sidereal time due to the shadow of the Earth

The results obtained in the investigation of possible diurnal effects for low-energy *single-hit* scintillation events of DAMA/LIBRA–phase1 (1.04 ton × year exposure) have been analysed in terms of Earth Shadow Effect, a model-dependent effect that could be expected in case of DM candidates inducing just nuclear recoils and having high cross-section with ordinary matter, which implies low DM local density in order to fulfil the DAMA/LIBRA DM annual modulation results [56].

In fact a diurnal variation of the low energy rate could be expected [70, 71] for these specific candidates, because during the sidereal day the Earth shields a given detector with a variable thickness, eclipsing the wind of DM particles. The induced effect should be a daily variation of their velocity distribution, therefore of their flux and, of course, of the signal rate measured deep underground. However, this effect is very small and would be appreciable only in case of high cross-section spin independent coupled candidates that could constitute a little fraction (ξ) in the Galactic dark halo.

The Earth's velocity in the galactic frame, $\vec{v}_e(t)$[1], defines an angle, θ, with the vector joining the center of the Earth to the position of the laboratory. Because of the Earth's rotational motion, the θ angle varies with the diurnal sidereal time and ranges between a minimum and a maximum which depend on the laboratory position on the Earth. In particular the larger is the allowed range of θ the larger is the effect resulting. The time dependence of θ can be expressed as a function of the laboratory latitude, λ, according to $\cos\theta(\mathrm{t}) = \cos\psi\cos\lambda\cos(\omega\mathrm{t} + \phi_0) + \sin\psi\sin\lambda$, where $\omega = (2\pi/24^h)$ is the Earth's angular rotational velocity and t the sidereal time in hours. The angles $\psi = (48.1)^0$ and $\phi_0 = (42.4)^0$ depend on the orientation of the Earth axis in the galactic frame [72].

By the fact, the diurnal variation of the velocity distribution and of the counting rate proves undetectable for cross-sections on nucleon $\sigma_n \leq 10^{-3}$ pb (being the latter dependent on the DM particle mass m_{DM}), since the Earth is practically transparent to similar particles, while high cross-section values give velocity distributions (and counting rates) significantly dependent on the considered time interval, producing a well detectable diurnal effect.

A study on diurnal variation in the rate with suitable exposure and stability could allow to investigate in some model scenarios high cross sections DM particle component (with small ξ) in the dark halo and decouple ξ from σ_n.

In particular, for each set of parameters [56], one can evaluate the $\xi\sigma_n$ allowed values as:

$$\xi\sigma_n = \frac{S_m^{exp}}{S'_{m,(2-4)keV}(m_{DM}, \sigma_n)}. \tag{3}$$

This corresponds, once including the experimental uncertainties on S_m^{exp}, to a band in the ξ vs σ_n plane.

Finally, for each considered set of parameters the three-dimensional allowed region in the parameter's space: ξ, σ_n, m_{DM} can be studied including the positive results from the DM annual modulation analysis of the DAMA/LIBRA–phase1 data [46], as reported for example in Fig. 8. In this considered scenario for quenching factors Q_I with *channeling* effect [56], B parameters set, $v_0 = 220$ km/s and $m_{DM} = 60$ GeV, the obtained upper limits on ξ do exclude $\sigma_n > 0.05$ pb and $\xi > 10^{-3}$. When also including other uncertainties as other halo models, etc. the results would be extended [56].

[1] The Earth's orbital speed around the Sun, which is the most relevant component for the annual modulation signature among the terms contributing to $\vec{v}_e(t)$, can be neglected in the present case because the data, taken in different periods during the year, contribute approximately to the same diurnal time intervals.

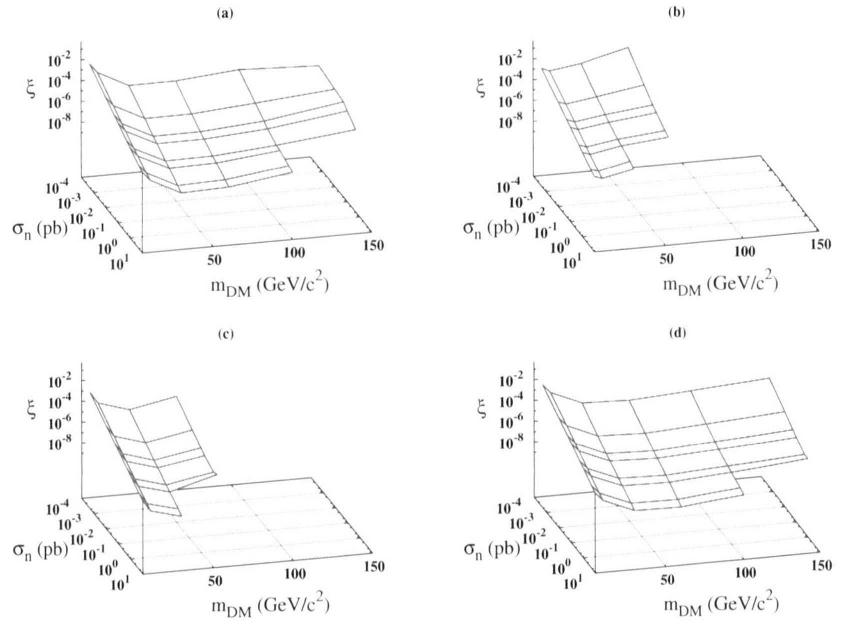

Figure 8. Examples of the mean values of the allowed region of ξ as function of σ_n and m_{DM}, represented as an allowed surface. The plots have been obtained for $v_0 = 220$ km/s by considering different scenarios regarding quenching factors value and other uncertainties [56]. We note that the "thickness" of the allowed regions around the surfaces is $\leq \pm 30\%$; therefore, for simplicity it is not represented in these figures. Finally, we recall that other uncertainties not considered here are present and can extend the result [56].

6.3. Directionality with anisotropic scintillators

The directionality approach – based on the study of the correlation between the recoil direction of the target nuclei and the Earth motion in the galactic rest frame – can offer a good approach to study those DM candidate particles able to induce just nuclear recoils. In fact, the dynamics of the rotation of the Milky Way galactic disc through the halo of DM causes the Earth to experience a wind of DM particles apparently flowing along a direction opposite to that of the solar motion relative to the DM halo. Hence, in the case of DM candidate particles interacting with nuclei the induced nuclear recoils are expected to be strongly correlated with the impinging direction of DM, while the background events are not; therefore, the study of the nuclear recoils direction can offer a way for pointing out the presence of these DM candidate particles.

In the practice, this approach has some technical difficulties because it is arduous to detect the short recoil track. Different techniques are under consideration but, up to now, they are at R&D stage and have not produced yet competitive results in the field (see e.g. DRIFT [73], DMTPC [74], DAMIC [75], or NEWS [76]). In fact, they are generally limited by the difficulty of detecting very short tracks and of achieving high stability, large sensitive volume and very good spatial resolution.

To overcome such a difficulty, it has been suggested the use of anisotropic scintillator detectors [77, 78, 79]; their use was proposed for the first time in Ref. [77] and revisited in Ref. [78]. In particular, low background $ZnWO_4$ crystal scintillators have been recently proposed since their features and performances are very promising [80]. In fact, both the light output and the scintillation pulse shape depend on the impinging direction of heavy particles (p, alpha, nuclear recoils, etc.) with respect to the crystal axes and can supply two independent ways to study the directionality and to discriminate the electromagnetic background (that does not give rise to any anisotropic effects).

Other advantages offered by $ZnWO_4$ detectors are very good radio-purity levels (about 0.1 cpd/kg/keV at low energy) and the potentiality to reach energy thresholds at keV level. Both these features can also be improved (e.g., the light yield shows a significant enhancement when working at low temperatures – about 100 K – and better radiopurity levels can be reached with dedicated R&D). A detailed discussion can be found in Ref. [80].

7. DAMA/LIBRA–phase2 and perspectives

After a first upgrade of the DAMA/LIBRA set-up in September 2008, a more important upgrade has been performed at the end of 2010 when all the PMTs have been replaced with new ones having higher Quantum Efficiency (Q.E.), realized with a special dedicated development by HAMAMATSU co.. Details on the developments and on the reached performances in the operative conditions are reported in Ref. [48]. We remind that up to October 2010 low background PMTs, developed by EMI-Electron Tubes with dedicated R&D, were used; the light yield and other response features already allowed in DAMA/NaI and DAMA/LIBRA–phase1 a software energy threshold of 2 keV in the data analysis. The feasibility to decrease the software energy threshold below 2 keV in the new configuration has been demonstrated [48].

Since the fulfillment of this upgrade, DAMA/LIBRA–phase2 – after optimization periods – is continuously running in order: (1) to increase the experimental sensitivity lowering the software energy threshold of the experiment; (2) to improve the corollary investigation on the nature of the DM particle and related astrophysical, nuclear and particle physics arguments; (3) to investigate other signal features and second order effects. This requires long and heavy full time dedicated work for reliable collection and analysis of very large exposures. DAMA/LIBRA also continue its study on several other rare processes [51, 52, 53, 54, 55, 56, 57].

Another upgrade at the end of 2012 was successfully concluded: new-concept preamplifiers were installed, with suitable operative and electronic features; in particular, they allow the direct connection of the signal to the relative channel of the Transient Digitizer (TD). Moreover, further improvements are planned; in particular, new trigger modules have been prepared and ready to be installed.

In the future DAMA/LIBRA will also continue its study on several other rare processes [51, 52, 53] as also the former DAMA/NaI apparatus did [81].

Finally, further improvements to increase the sensitivity of the set-up can be considered; in particular, the use of high Q.E. and ultra-low background PMTs directly coupled to the NaI(Tl) crystals is an interesting possibility[2]. This possible configuration can allow a further large improvement in the light collection and a further lowering of the software energy threshold. Moreover, efforts towards a possible highly radiopure NaI(Tl) "general purpose" experiment (DAMA/1ton) having full sensitive mass of 1 ton (we already proposed in 1996 as a general purpose set-up) have been continued in various aspects.

8. Conclusions

Large efforts are in progress with different approaches and target materials to investigate various kinds of DM candidates and scenarios. Due to the difficulty of measuring at very low energy, several techniques still would require further work for results' qualifications before enlarging their target mass.

[2] However, this would require the disassembling of the detectors since the light guides act at present also as optical windows.

As regards the possibility to exploit the directionality for some DM candidates, new efforts have to be encouraged towards a first realistic exploitation.

The DM model independent annual modulation signature with widely sensitive target materials still remains a major approach, offering an unique possibility for detection; it requires well known techniques, full proved detector stability, well known and proved detector response in all the aspects, etc. At present the DAMA positive model independent evidence for the presence of DM particles in the galactic halo is supported at very high confidence level. It has been shown in literature that this is compatible with many DM scenarios.

I have been also recalled the recent possible positive hints exploiting different approaches and different target materials, and the existing uncertainties in the model dependent results and comparisons.

Finally very useful complementary results can arise from experiment exploiting other target detectors and approaches adopting adequately safe experimental procedures.

9. Acknowledgements

It is a pleasure to thank the spokesperson, Prof. Rita Bernabei, and all the members of the DAMA Collaboration, without whom all this work would not have been possible. I am grateful to Conacyt for the support through the Mesoamerican Center for Theoretically Physics.

References

[1] R. Bernabei et al., *Int. J. Mod. Phys. A* **22**, 3155 (2007).
[2] R. Bernabei et al., *Eur. Phys. J. C* **23**, 61 (2002).
[3] S. Chang et al., *Phys. Rev. D* **79**, 043513 (2009).
[4] S. Chang et al., *Phys. Rev. Lett.* **106**, 011301 (2011).
[5] S. Chang et al., *J. Cosm. Astrop. Phys.* **08**, 018 (2010).
[6] D. Tucker-Smith and N. Weiner, *Phys. Rev. D* **64**, 043502 (2001).
[7] D. Tucker-Smith and N. Weiner, *Phys. Rev. D* **72**, 063509 (2005).
[8] R. Bernabei et al., *Mod. Phys. Lett. A* **23**, 2125 (2008).
[9] R. Bernabei et al., *Phys. Rev. D* **77**, 023506 (2008).
[10] R. Bernabei et al., *Int. J. Mod. Phys. A* **21**, 1445 (2006).
[11] R. Bernabei et al., in *Liquid Noble gases for Dark Matter searches: a synoptic survey* (Exorma Ed., Roma, ISBN 978-88-95688-12-1, 2009), pp. 1–53; arXiv:0806.0011v2.
[12] R. Bernabei et al., *Int. J. Mod. Phys. A* **30**, 1530053 (2015), 73 pages.
[13] Z. Ahmed et al., *Science* **327**, 1619 (2010).
[14] E. Armengaud et al., *Phys. Lett. B* **702**, 329 (2011).
[15] Z. Ahmed et al., arXiv:1203.1309.
[16] J.I. Collar and N.E. Fields, arXiv:1204.3559.
[17] R. Agnese et al., *Phys. Rev. Lett.* **112**, 041302 (2014).
[18] R. Agnese et al., arXiv:1509.02448.
[19] J. Sander et al., *AIP Conf. Proc.* **1534**, 129 (2013).
[20] R. Agnese et al., *Phys. Rev. Lett.* **112**, 241302 (2014).
[21] R. Agnese et al., *Phys. Rev. D* **88**, 031104(R) (2013).
[22] R. Agnese et al., *Phys. Rev. Lett.* **111**, 251301 (2013).
[23] T. de Boissiére, EDELWEISS coll., arXiv:1504.00820.
[24] S. Scorza, *J. of Phys.: Conf. Series* **718**, 042053 (2016).
[25] G. Angloher et al., *Eur. Phys. J. C* **72**, 1971 (2012).
[26] G. Angloher et al., *Eur. Phys. J. C* **74**, 3184 (2014).
[27] E. Aprile et al., *Phys. Rev. Lett.* **109**, 181301 (2012).
[28] J.I. Collar and D.N. McKinsey, arXiv:1005.0838; arXiv:1005.3723; J.I. Collar, arXiv:1006.2031; arXiv:1010.5187; arXiv:1103.3481; arXiv:1106.0653; arXiv:1106.3559.
[29] D.S. Akerib et al., *Phys. Rev. Lett.* **112**, 091303 (2014).
[30] D.S. Akerib et al., arXiv:1512.03506.
[31] P. Agnes et al., *Phys. Lett. B* **743**, 456 (2015).
[32] C.E. Aalseth et al., *Phys. Rev. Lett.* **106**, 131301 (2011).
[33] C.E. Aalseth et al., arXiv:1401.3295.

[34] J. Amaré et al., *J. of Phys.: Conf. Series* **718**, 042052 (2016)
[35] A.K. Drukier, K. Freese, and D.N. Spergel, *Phys. Rev. D* **33**, 3495 (1986); K. Freese, J. A. Frieman and A. Gould, *Phys. Rev. D* **37**, 3388 (1988).
[36] K. Freese et al., *Phys. Rev. D* **71**, 043516 (2005); K. Freese et al., *Phys. Rev. Lett.* **92**, 11301 (2004).
[37] R. Bernabei et al., *Il Nuovo Cim. A* **112**, 545 (1999).
[38] R. Bernabei et al., *Eur. Phys. J. C* **18**, 283 (2000).
[39] R. Bernabei el al., *La Rivista del Nuovo Cimento* **26** n.1, 1-73 (2003).
[40] R. Bernabei et al., *Int. J. Mod. Phys. D* **13**, 2127 (2004).
[41] R. Bernabei et al., *Eur. Phys. J. C* **47**, 263 (2006).
[42] R. Bernabei et al., *Eur. Phys. J. C* **53**, 205 (2008).
[43] R. Bernabei et al., *Nucl. Instr. and Meth. A* **592**, 297 (2008).
[44] R. Bernabei et al., *Eur. Phys. J. C* **56**, 333 (2008).
[45] R. Bernabei et al., *Eur. Phys. J. C* **67**, 39 (2010).
[46] R. Bernabei et al., *Eur. Phys. J. C* **73**, 2648 (2013).
[47] P. Belli et al., *Phys. Rev. D* **84**, 055014 (2011).
[48] R. Bernabei et al., *J. of Instr.* **7**, P03009 (2012).
[49] R. Bernabei et al., *Eur. Phys. J. C* **72**, 2064 (2012).
[50] R. Bernabei et al., *Int. J. of Mod. Phys. A* **28**, 1330022 (2013).
[51] R. Bernabei et al., *Eur. Phys. J. C* **62**, 327 (2009).
[52] R. Bernabei et al., *Eur. Phys. J. C* **72**, 1920 (2012).
[53] R. Bernabei et al., *Eur. Phys. J. A* **49**, 64 (2013).
[54] R. Bernabei et al., *Eur. Phys. J. C* **74**, 2827 (2014).
[55] R. Bernabei et al., *Eur. Phys. J. C* **74**, 3196 (2014).
[56] R. Bernabei et al., *Eur. Phys. J. C* **75**, 239 (2015).
[57] A. Addazi et al, *Eur. Phys. J. C* **75**, 400 (2015).
[58] K. Freese et al., *Phys. Rev. D* **71**, 043516 (2005); *New Astr. Rev.* **49**, 193 (2005); astro-ph/0310334; astro-ph/0309279.
[59] G. Gelmini, P. Gondolo, *Phys. Rev. D* **64**, 023504 (2001).
[60] F.S. Ling, P. Sikivie and S. Wick, *Phys. Rev. D* **70**, 123503 (2004).
[61] R. Bernabei et al., *AIP Conf. Proceed.* **1223**, 50 (2010) [arXiv:0912.0660].
[62] R. Bernabei et al., *J. Phys.: Conf. Series* **203**, 012040 (2010) [arXiv:0912.4200]; *http://taup2009.lngs.infn.it/slides/jul3/nozzoli.pdf*, talk given by F. Nozzoli.
[63] R. Bernabei et al., in the volume *Frontier Objects in Astrophysics and Particle Physics*, ed. S.I.F. (Vulcano, 2010), p. 157 [arXiv:1007.0595].
[64] R. Bernabei et al., *Can. J. Phys.* **89**, 11 (2011).
[65] R. Bernabei et al., *Physics Procedia* **37**, 1095 (2012).
[66] R. Bernabei et al., arXiv:1210.6199.
[67] R. Bernabei et al., arXiv:1211.6346.
[68] S. Scopel et al., arXiv:1512.08577, to appear in the Proceed. of the 14th Marcel Grossmann Meeting.
[69] P. Belli, A. Incicchitti and F. Cappella, *Int. J. of Mod. Phys. A* **29**, 1443011 (2014).
[70] J.I. Collar et al., *Phys. Lett. B* **275**, 181 (1992).
[71] J.I. Collar and F.T. Avignone III, *Phys. Rev. D* **47**, 5238 (1993).
[72] K.R. Lang, Astrophysical Formulae (Springer-Verlag) 1980.
[73] J.B.R. Battat et al., arXiv:1410.7821.
[74] J.P. Lopez et al., *Nucl. Instrum. and Meth. A* **696**, 121 (2012).
[75] A. Aguilar-Arevalo et al., arXiv:1510.021026.
[76] A. Alexandrov et al., *JINST* **9**, C12053 (2014).
[77] P. Belli et al., *Il Nuovo Cim. C* **15**, 475 (1992).
[78] R. Bernabei et al., *Eur. Phys. J. C* **28**, 203 (2003).
[79] N.J.C. Spooner at al., Proceedings of Int. Workshop on *The Identification of Dark Matter (IDM 1996)*, 481 (World Scientific, Singapore, 1997); Y. Shimizu et al., *Nucl. Instrum. and Meth. A* **496**, 347 (2003).
[80] F. Cappella et al., *Eur. Phys. J. C* **73**, 2276 (2013).
[81] R. Bernabei et al., *Phys. Lett. B* **408** 439 (1997); P. Belli et al., *Phys. Lett. B* **460** 236 (1999); R. Bernabei et al., *Phys. Rev. Lett.* **83** 4918 (1999); P. Belli et al., *Phys. Rev. C* **60** 065501 (1999); R. Bernabei et al., *Il Nuovo Cimento A* **112** 1541 (1999); R. Bernabei et al., *Phys. Lett. B* **515** 6 (2001); F. Cappella et al., *Eur. Phys. J.-direct C* **14** 1 (2002); R. Bernabei et al., *Eur. Phys. J. A* **23** 7 (2005); R. Bernabei et al., *Eur. Phys. J. A* **24** 51 (2005); R. Bernabei et al., *Astrop. Phys.* **4** (1995).

IOP Conf. Series: Journal of Physics: Conf. Series **866** (2017) 012017 doi:10.1088/1742-6596/866/1/012017

Detection of EASs at high altitude with ARGO-YBJ

Giuseppe Di Sciascio

INFN - Sezione Roma Tor Vergata, Italy

E-mail: giuseppe.disciascio@roma2.infn.it

Abstract. The ARGO-YBJ experiment has been in stable data taking for about 5 years at the YangBaJing Cosmic Ray Observatory (Tibet, P.R. China, 4300 m a.s.l., 606 g/cm^2). With a duty-cycle greater than 86% the detector collected about 5×10^{11} events in a wide energy range, from few hundreds GeV up to about 10 PeV. Exploiting the full coverage approach with a high segmentation of the readout at high altitude, ARGO-YBJ imaged the front of Extensive Air Showers (EAS) with unprecedented resolution and detail. A number of important problems in galactic cosmic ray physics has been faced through different analyses. In this contribution we summarize the latest results in gamma-ray astronomy and in cosmic ray physics.

1. The ARGO-YBJ experiment

ARGO-YBJ is a multipurpose experiment consisting in a dense sampling air shower array with 93% sensitive area located at very high altitude and devoted to the integrated study of gamma rays and cosmic rays with an energy threshold of a few hundreds GeV (for a summary of the ARGO-YBJ results see [1]).

The detector, located at the Yangbajing Cosmic Ray Observatory (Tibet, PR China, 4300 m a.s.l., 606 g/cm^2), is constituted by a central carpet $\sim$74$\times$78 m^2, made of a single layer of resistive plate chambers (RPCs) with $\sim$93% of active area, enclosed by a guard ring partially instrumented ($\sim$20%) up to $\sim$100$\times$110 m^2. The apparatus has a modular structure, the basic data acquisition element being a cluster (5.7$\times$7.6 m^2), made of 12 RPCs (2.85$\times$1.23 m^2 each). Each chamber is read by 80 external strips of 6.75$\times$61.80 cm^2 (the spatial pixels), logically organized in 10 independent pads of 55.6$\times$61.8 cm^2 which represent the time pixels of the detector [2]. The readout of 18,360 pads and 146,880 strips is the experimental output of the detector. The relation between strip and pad multiplicity has been measured and found in fine agreement with the Monte Carlo prediction [2]. In addition, in order to extend the dynamical range up to PeV energies, each chamber is equipped with two large size pads (139$\times$123 cm^2) to collect the total charge developed by the particles hitting the detector [3]. The RPCs are operated in streamer mode by using a gas mixture (Ar 15%, Isobutane 10%, TetraFluoroEthane 75%) for high altitude operation [4]. The high voltage settled at 7.2 kV ensures an overall efficiency of about 96% [5]. The central carpet contains 130 clusters (hereafter ARGO-130) and the full detector is composed of 153 clusters for a total active surface of $\sim$6700 m^2 (Fig. 1). The total instrumented area is $\sim$11,000 m^2. The information on strip multiplicity and the arrival times recorded by each pad are received by a local station devoted to manage the data of each cluster. A central station collects the information of all the local stations. The time of each fired pad in a window of 2 μs around the trigger time and its location are used to reconstruct the position of the shower core and the arrival direction of the primary particle. In order to perform

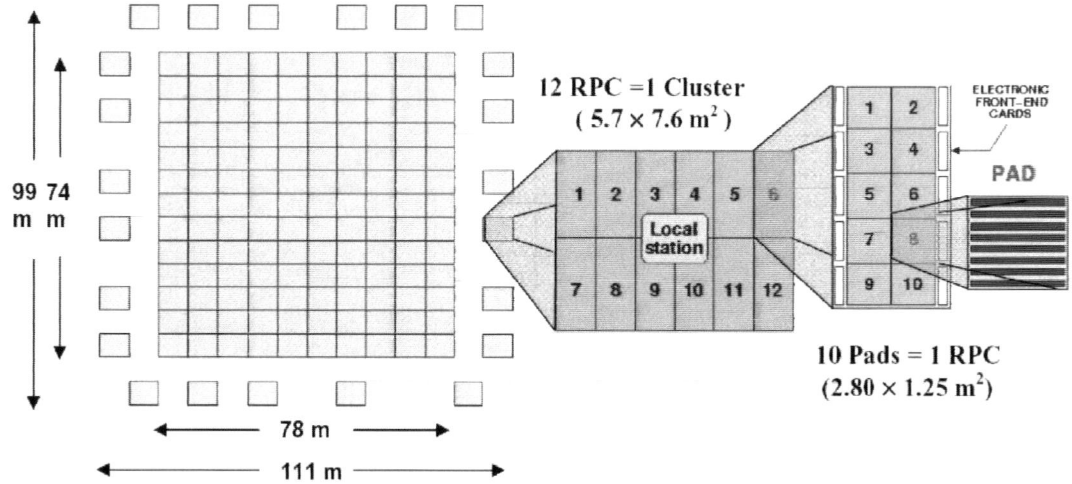

Figure 1. Layout of the ARGO-YBJ experiment (see text for a detailed description of the detector).

the time calibration of the 18,360 pads, a software method has been developed [6]. To check the stability of the apparatus a control system monitors continuously the current of each RPC, the gas mixture composition, the high voltage distribution as well as the environment conditions (temperature, atmospheric pressure, humidity).

The detector is connected to two different data acquisition systems, working independently, and corresponding to the two operation modes, shower and scaler. In shower mode, for each event the location and timing of every detected particle is recorded, allowing the reconstruction of the lateral distribution and the arrival direction [7, 8]. In scaler mode the total counts on each cluster are measured every 0.5 s, with limited information on both the space distribution and arrival direction of the detected particles, in order to lower the energy threshold down to $\sim$1 GeV [9].

In shower mode, a simple, yet powerful, electronic logic has been implemented to build an inclusive trigger. This logic is based on a time correlation between the pad signals depending on their relative distance. In this way, all the shower events giving a number of fired pads $N_{pad} \geq N_{trig}$ in the central carpet in a time window of 420 ns generate the trigger. This trigger can work with high efficiency down to N_{trig} = 20, keeping negligible the rate of random coincidences.

Because of the small pixel size, the detector is able to record events with a particle density exceeding 0.003 particles m^{-2}, keeping good linearity up to a core density of about 15 particles m^{-2}. This high granularity allows a complete and detailed three-dimensional reconstruction of the front of air showers at an energy threshold of a few hundred GeV, as can be appreciated in Fig. 2 (left plot) where a typical shower detected by ARGO-YBJ is shown. Showers induced by high energy primaries (>100 TeV) are also imaged by the charge readout of the large size pads (see Fig. 2, right plot), which allow the study of shower core region with unprecedented resolution [3].

The whole system, in smooth data taking since July 2006 with ARGO-130, has been in stable data taking with the full apparatus of 153 clusters from November 2007 to January 2013, with the trigger condition N_{trig} = 20 and a duty cycle $\geq$86%. The trigger rate is $\sim$3.5 kHz with a

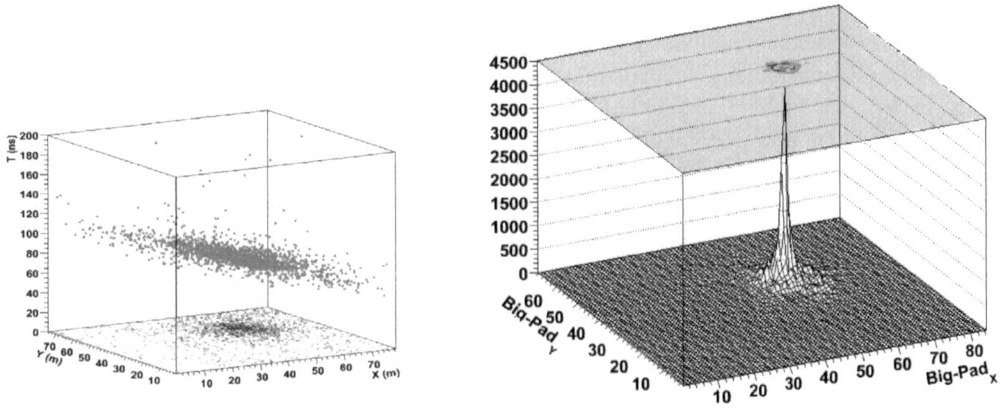

Figure 2. Showers imaged by the ARGO-YBJ central carpet with the digital (left plot: the space-time structure of a low energy shower provided by the strip/pad system) and the analog (right plot: the core of a high energy shower imaged by the charge read-out) read-outs.

dead time of 4%.

1.1. Event reconstruction and data selection in shower mode

The reconstruction of the shower parameters is carried out through the following steps.

At first, a plane surface is analytically fitted (with weights equal to 1) to the shower front. This procedure is repeated up to 5 times, each iteration rejecting hits whose arrival time is farther than 2 standard deviations from the mean of the distribution of the time residuals from the fitted plane surface. This iterative procedure is able to reject definitively from the reconstruction the time values belonging to the non-Gaussian tails of the arrival time distribution [10]. After this first step the problem is reduced to the nearly-vertical case by means of a projection which makes the fit plane overlapping the detector plane. Thereafter, the core position, i.e. the point where the shower axis intersects the detection plane, is obtained fitting the lateral density distribution of the secondary particles to a modified Nishimura-Kamata-Greisen (NKG) function. The fit procedure is carried out via the maximum likelihood method [7]. Finally, the core position is assumed to be the apex of a conical surface to be fitted to the shower front. The slope of such a conical correction is fixed to $\alpha = 0.03$ ns/m [10].

The capability of reconstructing the primary arrival direction can be further enhanced by applying robust statistical methods in the analysis of the shower front, conveniently weighting the contribution of the most delayed particles. In detail, we first fit a conical surface to the shower image, by minimizing the sum of the squares of the time residuals. At this stage, all the particles hitting the detector have the same weight w_i=1. After computing the RMS of the time residual distribution with respect to such a conical surface, we set K = 2.5·RMS as a 'scale parameter' and perform the minimization of the square of the time residuals weighted sum, where w_i=1 if the particle is onward the shower front, $w_i = f((t_i^{exp} - t_i^{fit})/K)$ otherwise. The function $f(x)$ is a common Tukey biweight function [11]. The fit procedure is iterated, every time refreshing the scale parameter, until the last reconstructed direction differs from the previous one for less than 0.1°.

1.2. Scaler Mode Technique

The study of transient phenomena (such as GRBs, solar flares) can be successfully performed by ground-based experiments such as EAS detectors down to about 1 GeV working in "single particle mode", i.e. counting all the particles hitting the individual detectors, independently of whether they belong to a large shower or they are the lone survivors of small showers [12] . Because of the cosmic ray spectrum steepness, most of the events detected with this technique are in fact due to solitary particles from small showers generated by 1–100 GeV cosmic rays.

The power of this technique is in its extreme simplicity: it is sufficient to count all the particles hitting the detectors during fixed time intervals (i.e. every second or more frequently, depending on the desired time resolution) and to study the counting rate behaviour versus time, searching for significant increases. The observation of an excess in coincidence with a transient phenomenon detected by a satellite would be an unambiguous signature of the nature of the signal. The single particle technique does not allow one to measure energy and direction of the primary particle, because the number of detected particles (often only one per shower) is too small to reconstruct the shower parameters. For this technique, an accurate knowledge of the detector and its sensitivity to both environmental and instrumental parameters is of crucial importance in order to correctly evaluate the statistical significance of the detected signals.

In ARGO-YBJ, the counting rate of each cluster is measured every 0.5 s, with no measurement of the lateral distribution and arrival direction of the detected particles. For each cluster the signal coming from the 120 pads is added up and put in coincidence in a narrow time window (150 ns), giving the rate of counts ≥ 1, ≥ 2, ≥ 3, ≥ 4, read by four independent scaler channels. The corresponding measured rates are, respectively, $\sim$40 kHz, $\sim$2 kHz, $\sim$300 Hz and $\sim$120 Hz for each cluster. The counting rates for a given multiplicity are then obtained with the relation $n_i = n_{\geq i} - n_{\geq i+1}$ for i = 1, 2, 3.

The study of the counting distribution for each cluster is important in order to monitor the stability of the detector and its statistical (Poissonian) behaviour. Fig. 3 shows the

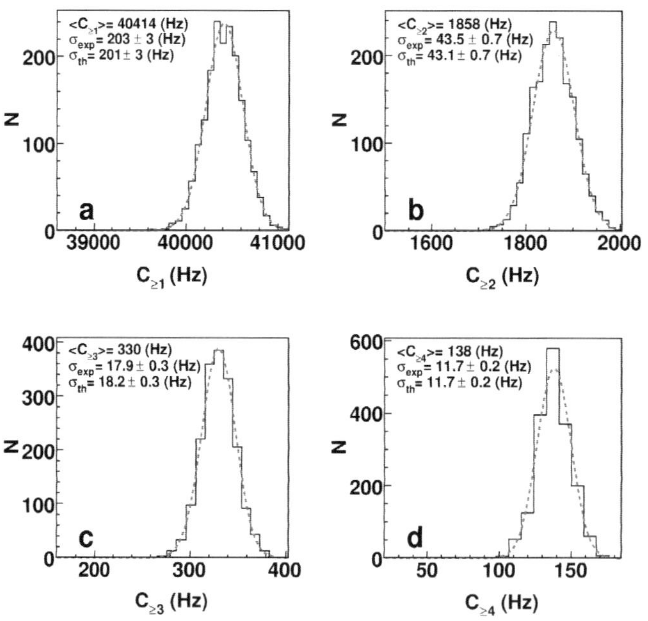

Figure 3. Experimental distributions of the counting rates and their Gaussian fits for a typical cluster: (a) $C_{\geq 1}$, (b) $C_{\geq 2}$, (c) $C_{\geq 3}$ and (d) $C_{\geq 4}$ for 30 minutes data accumulation. The standard deviation of the Gaussian fit (σ_{exp}) is compared with the square root of the mean of the experimental distribution (σ_{th}) to check the compatibility with the Poisson distribution.

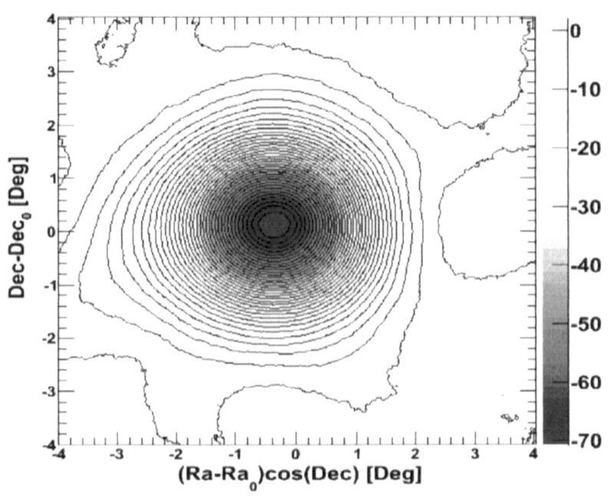

Figure 4. Significance map of the Moon region for events with N_{strip} >40, observed by ARGO-YBJ in the period July 2006 - November 2010. The coordinates are right ascension and declination centered on the Moon position. The color scale gives the statistical significance in terms of standard deviations.

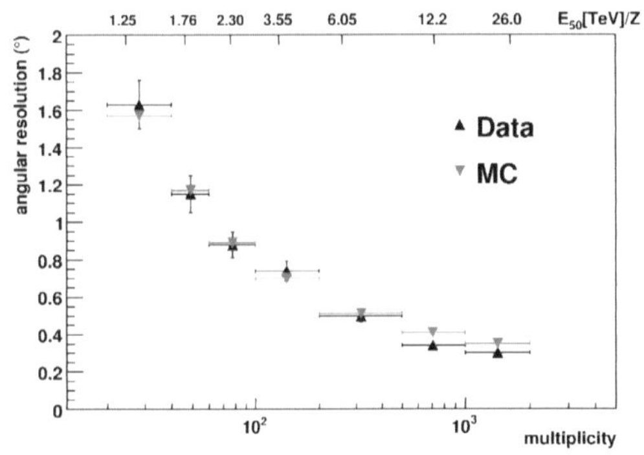

Figure 5. Measured angular resolution of the ARGO-YBJ detector compared to the expectations from MC simulation as a function of the particle multiplicity, i.e., the number of fired strips. The multiplicity bins are shown by the horizontal bars. The upper scale refers to the rigidity (TeV/Z) associated to the median energy in each multiplicity bin.

counting rates for multiplicities ≥1, ≥2, ≥3, ≥4 for a typical cluster during 30 minutes of data acquisition. The distributions for all $C_{\geq i}$ are Poissonian, meaning that the environmental parameter variations have negligible effects over such a time scale. Detailed discussions can be found in [9].

1.3. Angular resolution and energy scale calibration

The performance of the detector and the operation stability are continuously monitored by observing the Moon shadow, i.e., the deficit of cosmic rays detected in its direction. Indeed, the size of the deficit allows the measurement of the angular resolution and its position allows the evaluation of the absolute pointing accuracy of the detector. In addition, since charged particles are deflected by the geomagnetic field (GMF), the observation of the displacement of the shadow of the Moon provides a direct calibration of the relation between shower size (the recorded pad/strip multiplicity) and primary energy. The calibration of the absolute energy scale of ground-based detectors is one of the main problems affecting indirect measurements of cosmic rays.

ARGO-YBJ observes the Moon shadow with a sensitivity of about 9 standard deviations (s.d.) per month for events with a multiplicity N_{strip} ≥40 and zenith angle θ <50°, corresponding to a proton median energy E_p ~1.8 TeV. The significance map of the Moon region obtained with all data from July 2006 to November 2010 is shown in Fig. 4. The CR Moon shadowing effect is observed with a significance of about 70 s.d. . The data analysis and a full account of the results are described in [14]. The measured angular resolution is better than 0.5° for CR-induced showers with energies E > 5 TeV (Fig. 5), in excellent agreement with the MC evaluation. With the same simulation codes we find that the angular resolution for γ-rays is smaller by about 30–40%, depending on the multiplicity, due to the better defined time profile of the showers.

In Fig. 6 the displacement of the Moon shadow in the East-West direction as a function of the particle multiplicity, i.e. the number of fired strips N_{strip} on ARGO-YBJ central carpet, is shown. The observed westward shift is compared to the MC simulation of the cosmic ray primary propagation in Earth-Moon system [15]. The rigidity scale refers to the rigidity (TeV/Z) associated to the median energy in each multiplicity bin. The good agreement between data and simulation allows the attribution of this displacement to the combined effect of the detector point spread function (PSF) and the GMF, making us confident in the energy calibration of the detector. Therefore, the rigidity scale can be fixed in the multiplicity range 20-2000 particles, where the Moon shadow is moving under the bending effect of the GMF. The absolute rigidity scale uncertainty in the ARGO-YBJ experiment is estimated to be smaller than 13% in the energy range from 1 to 30 (TeV/Z) [14].

The displacement of the center of the Moon shadow in the North-South direction enables us to estimate the systematic error in pointing accuracy and its long-term stability aside from Monte Carlo simulations, since the East-West component of the GMF is almost zero at YangBaJing. The amount of CR deficit due to the Moon provides a good estimation of the size of the shadow, therefore of the angular resolution. The position of the Moon shadow measured with the ARGO-YBJ experiment turned out to be stable at level of 0.1° and the angular resolution is stable at level of 10%, on a monthly basis [14].

2. ARGO-YBJ recent highlights in Gamma-Ray Astronomy

2.1. The Fermi Cocoon in the Cygnus region

The Cygnus region is the brightest portion of the gamma ray northern sky, where several complex structures have been observed at different wavelengths. This region is rich in potential cosmic ray acceleration sites, including Wolf-Rayet stars, OB associations, supernova remnants, pulsars, microquasars and superbubbles. Various Very High Energy (VHE) gamma ray sources have been detected within the Cygnus region, the brightest ones being the extended MGRO J2031+41/TeV J2032+4130 and MGRO J2019+37 [16], whose locations are consistent with two Fermi pulsars [17].

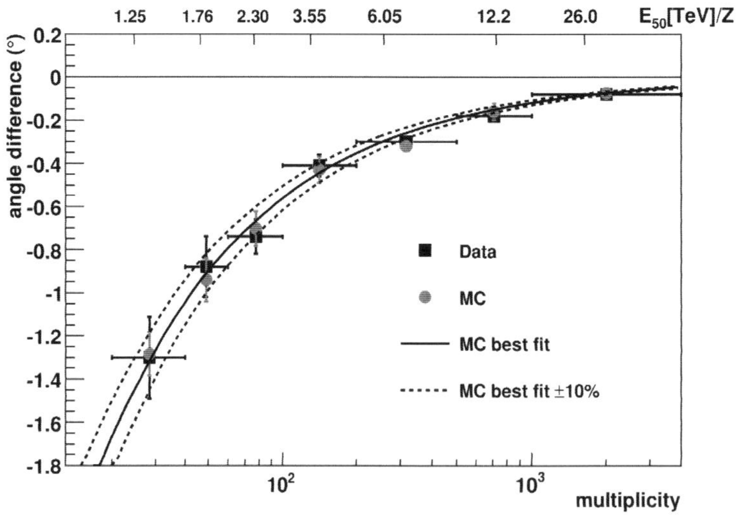

Figure 6. Measured westward displacement of the Moon shadow as a function of multiplicity (black squares). The data are compared to MC simulation (red circles). The upper scale refers to the median energy of rigidity (TeV/Z) in each multiplicity bin.

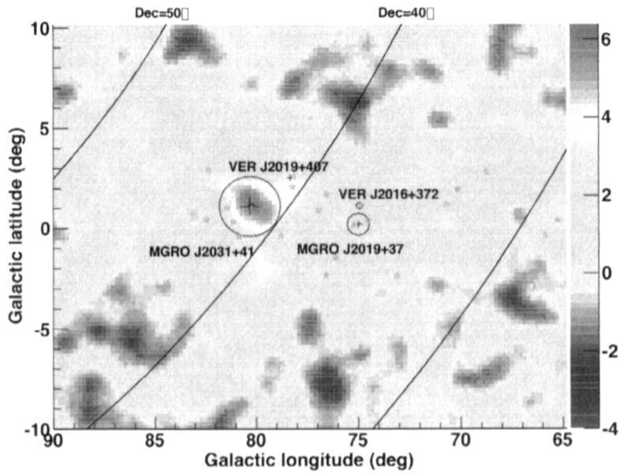

Figure 7. Significance map of the Cygnus region as observed by ARGO-YBJ. The four known VHE sources are reported. The errors on the MGRO source positions are given with crosses, while the circles indicate their intrinsic sizes. The open stars mark the location of the 24 GeV sources in the second Fermi/LAT catalogue.

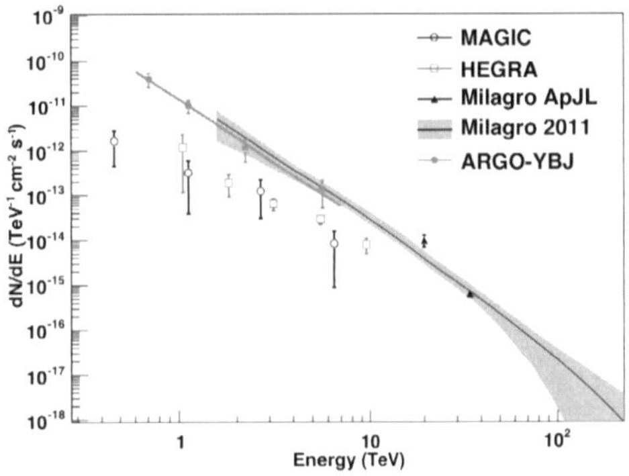

Figure 8. Spectrum of MGRO J2031+41 as measured by ARGO-YBJ and other experiments.

The significance map of the Cygnus region as observed by ARGO-YBJ using events with $N_{pad} > 20$ is shown in Figure 7 [18]. The four known TeV sources and the 24 GeV sources in the second Fermi/LAT catalogue are also marked in the figure. An excess is observed over a large part of the region, indicating a possible diffuse gamma ray emission (see next section).

A signal with a significance larger than 6 s.d. is detected at the position of MGRO J2031+41. The source extension is determined to be $\sigma_{ext} = (0.2^{+0.4}_{-0.2})^\circ$, consistent with the previous estimations by HEGRA [19] and MAGIC [20] experiments. Assuming an intrinsic extension $\sigma_{ext} = 0.1^\circ$, the energy spectrum measured in the range 0.6-7 TeV is $dN/dE = (1.40 \pm 0.34) \times 10^{-11}(E/1\ TeV)^{(-2.83\pm0.37)}$ photons cm^{-2} s^{-1} TeV^{-1} and is reported in Figure 8. The integral flux above 1 TeV is ~0.3 Crab unit, in agreement with Milagro results [21, 22] but about a factor 10 higher than the fluxes determined by HEGRA and MAGIC.

In a location consistent with MGRO J2031+41, Fermi/LAT detected a complex extended source, attributed to the emission by a *"cocoon"* of freshly accelerated cosmic rays which fill the cavities carved by stellar winds and ionization fronts from young stellar clusters [23]. After reanalysing the complete ARGO-YBJ data set, subtracting the contribution of the overlapping TeV sources and using a larger region to evaluate the excess map (since Fermi/LAT observations revealed a large extended source), ARGO J2031+4157 resulted with an extension $\sigma_{ext} = 1.8^\circ \pm 0.5^\circ$, consistent with that of the Cygnus Cocoon as measured by Fermi/LAT, i.e., $\sigma_{ext} = 2.0^\circ \pm 0.2^\circ$ [24]. The ARGO-YBJ view of the Cygnus Cocoon region (for $N_{pad} \geq 20$) is given in Figure 9, where the largest statistical significance is 6.1 s.d., at the position of ARGO J2031+4157.

The spectrum also shows a good connection with that determined by Fermi/LAT in the 1-100 GeV energy range, as shown in Figure 10. Therefore, ARGO J2031+4157 is identified as the counterpart of the Cygnus Cocoon at TeV energies. The combined differential spectrum of Fermi/LAT and ARGO-YBJ data is fitted with the power law function $dN/dE = (3.5 \pm 0.3) \times 10^{-9}(E/0.1\ TeV)^{(-2.16\pm0.04)}$ photons cm^{-2} s^{-1} TeV^{-1} (dot-dashed line in Figure 10).

In order to reproduce the gamma ray emission from the Cygnus Cocoon, a purely hadronic

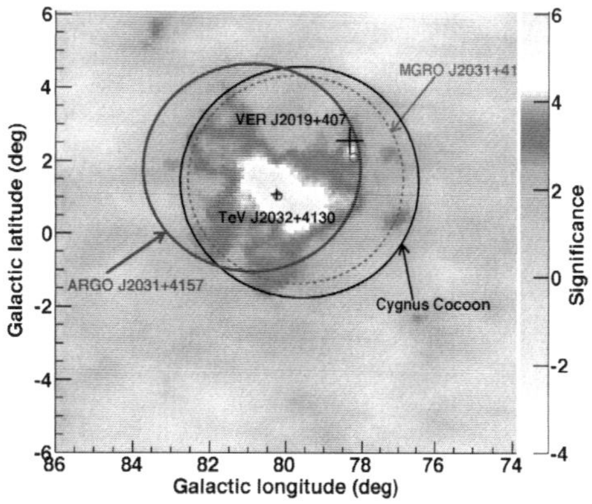

Figure 9. Significance map around ARGO J2031+4157 as observed by ARGO-YBJ. The large circles indicate the positions and 68% contours of ARGO J2031+4157, MGRO J2031+41 and the Cygnus Cocoon. The positions and extensions of TeV J2032+4130 [19] and VER J2019+407 [25] are marked with crosses.

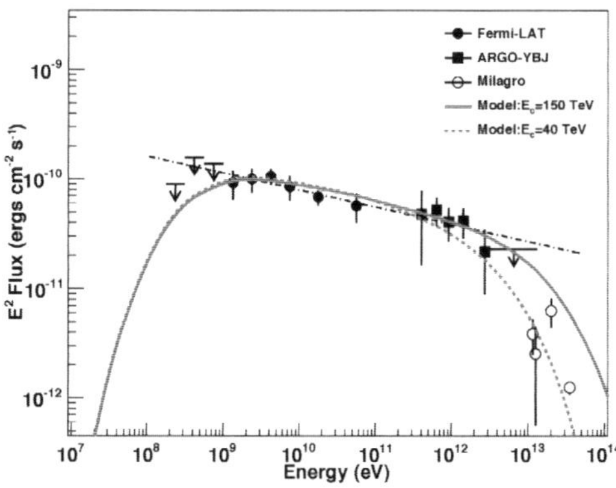

Figure 10. Spectrum of the Cygnus Cocoon as measured by different detectors. The arrows below 1 GeV indicate the upper limits set by Fermi/LAT, the Milagro data refer to MGRO J2031+41, at 12 TeV corrected for the extrapolation of TeV J2032+4130 [23]. The dot-dashed line shows the power law best fit to the combined Fermi/LAT and ARGO-YBJ data. The solid and dotted curves are predicted by a purely hadronic model with proton cutoff energy at 150 TeV and 40 TeV, respectively.

model [26] can be adopted, in which the observed gamma rays are due to the decay of π^0 mesons resulting from inelastic collisions between accelerated protons and target gas. Assuming that the primary proton spectrum follows a power law with exponential cutoff, the maximum cutoff energy allowed by the ARGO-YBJ highest energy upper limit is E_c=150 TeV (solid line in Figure 10). Taking into account also the Milagro data, E_c would be around 40 TeV (dotted line).

Concerning MGRO J2019+37, which is the second brightest source after the Crab Nebula with Milagro data at ∼12 TeV, having an extension $\sigma_{ext} = 0.32° \pm 0.12°$ [27], the ARGO-YBJ map does not show any excess above 3 s.d., and flux upper limits were set at 90% confidence level (c.l.) [18]. The VERITAS observatory imaged MGRO J2019+37, resolving it into two different sources: the faint point-like VER J2016+371 and the bright extended (∼ 1°) VER J2019+368, which likely contributes to the bulk of the emission observed by Milagro and is spatially coincident with the pulsar PSR J2021+3651 and the star formation HII region Sh 2-104 [28]. The flux of VER J2019+368 is in agreement with the ARGO-YBJ upper limit. All these results are reported in Figure 11.

Considering also the ARGO-YBJ results for MGRO J1908+06 [29] and HESS J1841-055 [30], as for the Milagro air shower detector, the fluxes measured in extended sources are systematically larger than those measured with Cherenkov telescopes. The origin of this discrepancy, which is not present for point-like sources, is not clear. For ARGO-YBJ, the overall systematic error on the flux measurements has been estimated to be < 30% [31]. There could be some systematic effect related to the methods for background evaluation of the different experimental techniques [30].

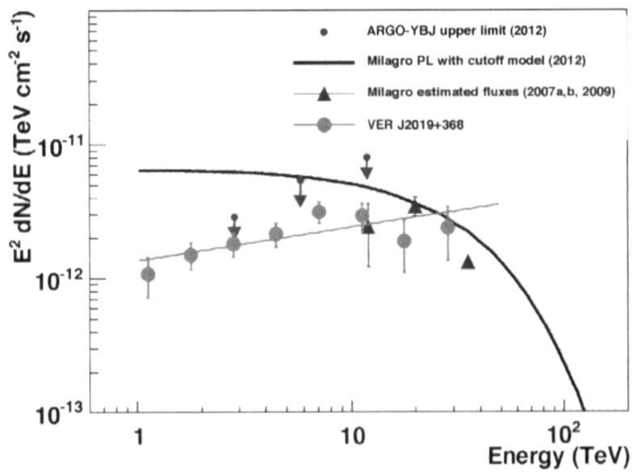

Figure 11. Observational results for MGRO J2019+37 from different experiments (figure taken from [28]). The Milagro power law with cutoff best fit, with its 1σ error band (shadowed area), is taken from [22].

A contribution from the diffuse gamma ray emission, according to ARGO-YBJ measurements, it is different for each extended source but always less than 15% of the detected flux [32].

2.2. Diffuse Gamma-Ray emission from the Galactic Plane

Diffuse gamma rays are produced by relativistic electrons by bremsstrahlung or inverse Compton scattering on background radiation fields, or by protons and nuclei via the decay of π^0 produced in hadronic interactions with interstellar gas. Thus, the space distribution of this emission can trace the location of the cosmic ray sources and the distribution of interstellar gas. On the other hand, the diffuse emission provides a background in the search for point sources.

In the GeV energy range, the EGRET data showed a significant excess compared to expectations (the 'EGRET GeV excess')[33] likely due to instrumental effects [34] because not confirmed by the more recent and accurate Fermi-LAT data [35, 36]. A diffuse gamma ray flux at TeV energies from the $30° < l < 110°$ longitudine range of the Galactic plane has been reported by Milagro [37, 38, 39].

The measured flux, once connected to the EGRET data by a power law with differential spectral index of −2.61, reveals a *"TeV excess"* in the diffuse gamma ray spectrum, being this flux 5–10 times higher than expectations [40]. However, as will be discussed in the following, the Milagro result does not take into account the contributions from the Cygnus cocoon and from the overlapping point or extended sources TeV J2032+4130, VER J2019+407 and VER J2016+372.

The events collected by ARGO-YBJ have been analysed to determine the diffuse gamma ray emission in the Galactic plane at Galactic longitudes $25° < l < 100°$ and Galactic latitudes $|b| < 5°$ [32]. This analysis was carried out in the energy range from ∼350 GeV to ∼2 TeV, connecting the region explored by Fermi/LAT with that investigated by Milagro. In particular, the analysis was focused on two selected regions of the Galactic plane, i.e., $40° < l < 100°$ and $65° < l < 85°$ (the Cygnus region), where Milagro observed an excess with respect to what predicted by current models.

In the Galaxy region $40° < l < 100°$, $|b| < 5°$, after masking the discrete sources and subtracting the residual contribution, an excess with a statistical significance of 6.1 s.d. above the background is found. The spectral analysis provides the three fluxes shown in Figure 12, at median energies 350 GeV, 680 GeV and 1.47 TeV (with uncertainties of about 30%). The fit to the ARGO-YBJ data with a power law gives a spectral index -2.90 ± 0.31, and the corresponding

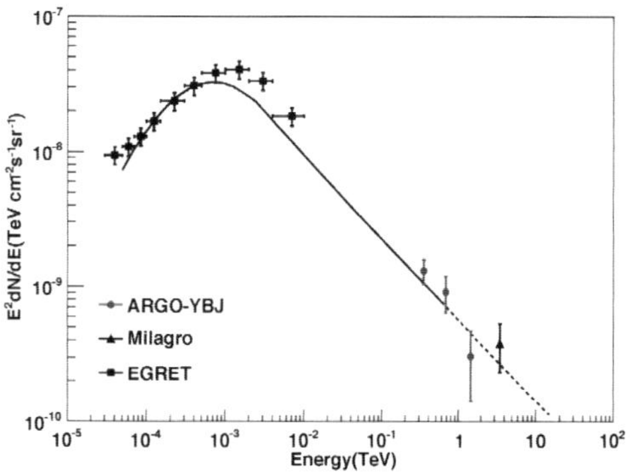

Figure 12. Energy spectrum of the diffuse gamma ray emission measured in the Galactic region $40° < l < 100°$, $|b| < 5°$ [32]. The line indicates the energy spectrum expected from the Fermi/LAT template (with spectral index -2.6, which also rules its short-dashed extension) [43].

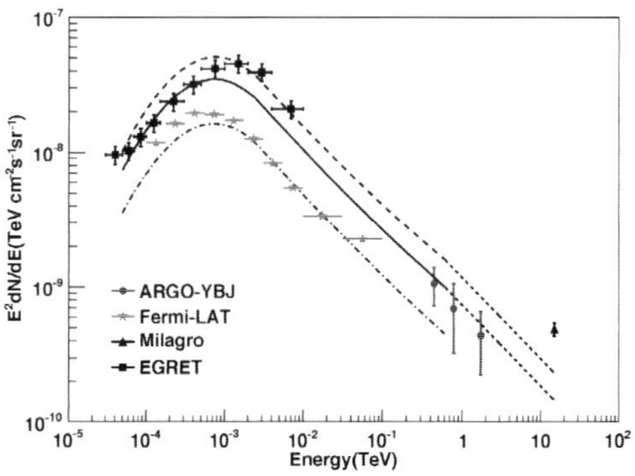

Figure 13. Energy spectrum of the diffuse gamma ray emission measured by ARGO-YBJ and EGRET in the Galactic region $65° < l < 85°$, $|b| < 5°$, by Milagro in the region $65° < l < 85°$, $|b| < 2°$ and by Fermi/LAT in $72° < l < 88°$, $|b| < 15°$ [32]. The different lines indicate the energy spectra expected from the Fermi/LAT template (with spectral index -2.6, which also rules their short-dashed extensions) in the different sky regions investigated by the detectors [43].

flux at 1 TeV is compatible with the extrapolation of the Fermi/LAT template [43].

Table 1. Diffuse gamma-ray emission from the Galactic plane for $|b| < 5°$.

Locations	Statistical Significance	Spectral Index
$25° < l < 100°$	6.9 s.d.	-2.80 ± 0.26
$40° < l < 100°$	6.1 s.d.	-2.90 ± 0.31
$65° < l < 85°$	4.1 s.d.	-2.65 ± 0.44
$25° < l < 65°$ & $85° < l < 100°$	5.6 s.d.	-2.89 ± 0.33

On the other hand, as mentioned, the first measurement of the diffuse TeV (integral) flux from the Galactic plane made by Milagro [41] revealed a "TeV excess" in the diffuse gamma ray spectrum with respect to expectations [42]. This Milagro measurement, converted into differential flux, is only 34% greater than the value expected from the extrapolation of the Fermi/LAT template, and within the experimental uncertainties (see triangle with error bars in Figure 12). Moreover, considering that the Milagro result does not take into account the contributions from the Cygnus Cocoon (not yet discovered at the time of the measurement) and from overlapping point and extended sources, the discrepancy with the Fermi/LAT predictions is almost cancelled out. Therefore, the full set of measurements with ground-based detectors is in agreement with direct observations by Fermi/LAT, and the evidence of any "TeV excess", requiring additional sources or particle production processes other than those producing Galactic cosmic rays, is ruled out.

In the Galactic region $65° < l < 85°$, $|b| < 5°$, after masking the discrete sources and the Cygnus Cocoon and subtracting the residual contribution, an excess of 4.1 s.d. is left. This direction points into our spiral arm at the remarkable Cygnus star-forming region, located at a distance of about 1.4 kpc. The spectral energy distribution of gamma ray emission is shown in Figure 13 (filled stars) [32] together with the spectrum expected from the Fermi/LAT template (dot-dashed line) [43]. Milagro measured the diffuse gamma ray emission from the region $65° < l < 85°$, $|b| < 2°$ at a median energy of 15 TeV [39], obtaining the flux reported as a filled triangle in Figure 13. For comparison, the long-dashed line shows the expected energy spectrum for this region according to the Fermi/LAT template. The Milagro flux results about 75% higher than the Fermi template, suggesting the presence of an excess. The spectral analysis of ARGO-YBJ data provides the three fluxes shown in Figure 13 at median energies 440 GeV, 780 GeV and 1.73 TeV (with uncertainties of about 40%). The fit to ARGO-YBJ data with a power law gives a spectral index -2.65 ± 0.44, and the corresponding flux at 1 TeV is about 10% lower than the extrapolation of the Fermi/LAT template. These data do not show any excess at energies around 1 TeV which corresponds to the excess found by Milagro at a median energy of 15 TeV. Again, this discrepancy can be explained taking into account that the contribution of all the discrete gamma ray sources was not completely removed from the Milagro data. According to the ARGO-YBJ data, the 1 TeV flux associated to the Cygnus Cocoon is of the same order of the diffuse flux.

The spectral indexes measured by ARGO-YBJ in different regions of the Galactic plane are reported in Table 1 [32]. The statistical significance of the observations in standard deviations (s.d.) is shown in the second column. The TeV flux averaged over the Cygnus region $65° < l < 85°$ cover the energy range 400 GeV – 2 TeV and follow a power law with spectral index -2.65 ± 0.44, a value very close to that found for TeV emission from the Cygnus Cocoon, indicating the presence of young cosmic rays accelerated by a nearby source.

2.3. 4.5 years observations of the blazar Mrk 421

Active Galactic Nuclei (AGNs), one of the most luminous sources of electromagnetic radiation in the universe, are galaxies with a strong and variable non-thermal emission, believed to be the result of accretion of mass onto a supermassive black hole (with a mass ranging from $\sim 10^6$ to $\sim 10^{10}$ $M_\odot$) lying at the center of the galaxy.

To understand the emission variability and the underlying acceleration and radiation mechanisms in jets, continuous multi-wavelength observations, particularly in X-rays and VHE γ-rays, are crucial. A simultaneous Spectral Energy Distribution (SED) could provide a snapshot of the emitting population of particles and also constrain the model parameters at a given time [44, 45]. The shape of particle energy distribution could bring information on the underlying acceleration processes (e.g.[46, 47, 48, 49]). In the VHE band, Cherenkov telescopes cannot regularly monitor AGNs, because of their limited duty cycle and narrow field of view. Wide-FOV EAS arrays, with high duty cycles, are more suitable for this purpose.

During 5 years ARGO-YBJ continuously monitored the blazar Mrk 421, extending at higher energies the multi-wavelength survey carried out by the Owens Valley Radio Observatory (OVRO), the satellite-borne X-ray detectors Swift, the Rossi X-ray Timing Explorer (RXTE), the Monitor of All-sky X-ray Image (MAXI) and the GeV γ-ray detector Fermi-LAT. In particular, thanks to the ARGO-YBJ and Fermi-LAT data, the high energy component of Mrk 421 SED has been completely covered without any gap from 100 MeV to 10 TeV from August 2008 to February 2013, a period that includes several large flares of Mrk 421. Such a long-term multi-wavelength observation is rare and provides a unique opportunity to investigate the emission variability of Mrk 421 from radio frequencies to TeV γ-rays ([50] and references therein).

Fig. 14 shows the light curves of Mrk 421, as obtained by the data of different experiments,

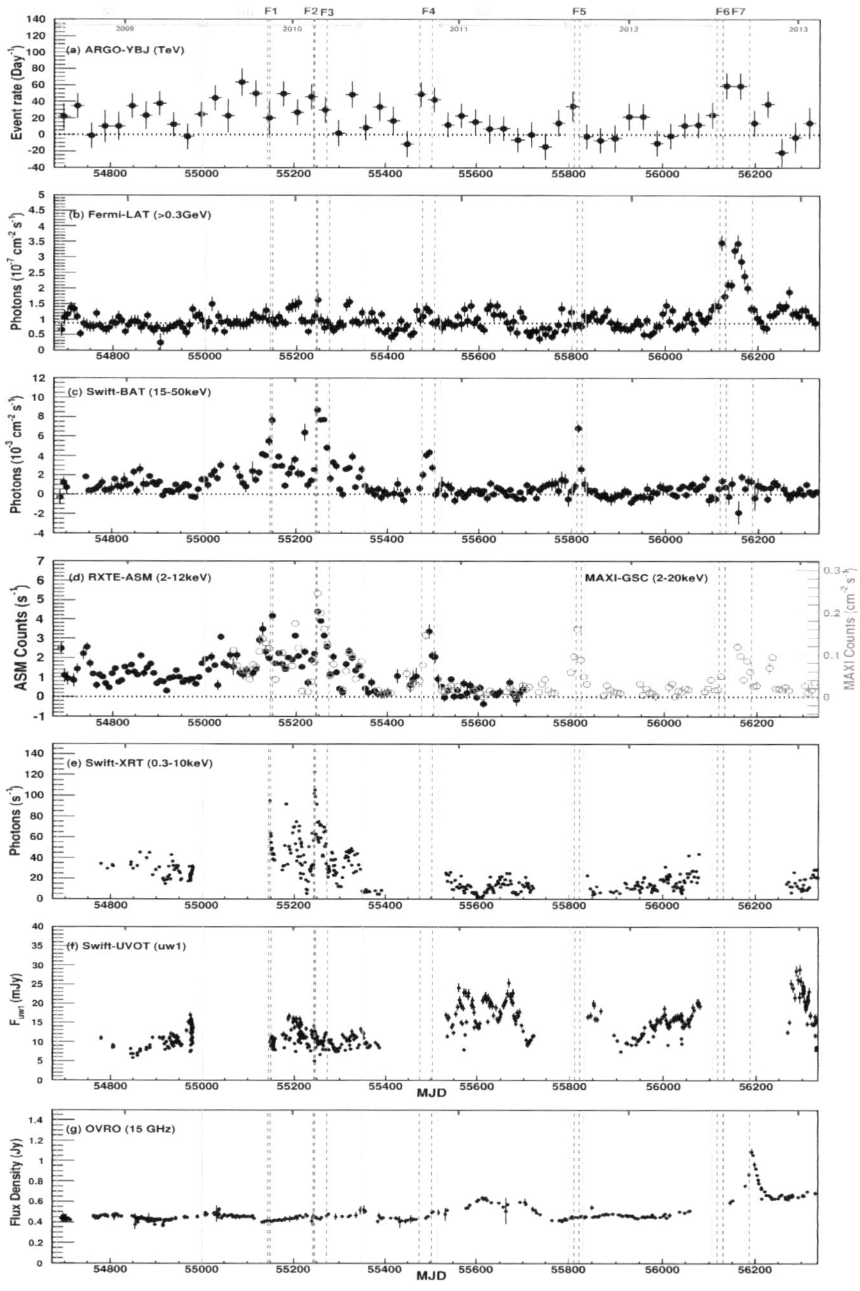

Figure 14. Mrk 421 light curves in different energy bands, from 2008 August 5 to 2013 February 7 [50]. Each bin of the ARGO-YBJ data contains the event rate averaged over 30 days.

covering the energy range from radio to the TeV band [50]. The time integration is chosen taking into account the sensitivity of the instruments. For ARGO-YBJ each point corresponds to one month (30 days) of data, while for Fermi-LAT, Swift-BAT, RXTE-ASM and MAXI-GSC the data are averaged over one week (see [50] and references therein). Thanks to the ARGO-YBJ and Fermi data, the whole energy range from 100 MeV to 10 TeV is covered without any gap. The main results of the multiwavelenght analysis can be summarized as follows [50]:

- Mrk 421 showed both low and high activity phases at all wavebands during the 4.5 year period analyzed. The variability increases with energy for both the SED components. Concerning the synchrotron component, the variability amplitude increases from 21% in radio and 33% in UV to 71%-73% in soft X-rays and 103%-137% in hard X-rays. For the Inverse Compton component, the amplitude is 39% at GeV energies and increases to 84% at TeV energies.
- The variation of the X-ray flux is clearly correlated with the TeV γ-ray flux. This result is consistent with many previous observations, supporting the idea that the X-ray and VHE γ-ray emissions originate from the same zone. For the first time we observed a moderate correlation between GeV and TeV γ-ray fluxes. On the contrary, X-ray and γ-ray fluxes are weakly or not correlated with radio and UV fluxes.
- Seven large flares, including five X-ray flares and two GeV γ-ray flares with variable durations (3$-$58 days), and one X-ray outburst phase were identified and used to investigate the variation of the spectral energy distribution with respect to a relative quiescent phase.
- During the outburst phase and the seven flaring episodes, the peak energy in X-rays is observed to increase from sub-keV to few keV. The TeV γ-ray flux increases up to 0.9$-$7.2 times the flux of the Crab Nebula. The behavior of GeV γ-rays is found to vary depending on the flare, a feature that leads us to classify flares into three groups according to the GeV flux variation.
- The one-zone synchrotron self-Compton (SSC) model was adopted to describe the emission spectra. Two out of three groups can be satisfactorily described using injected electrons with a power-law spectral index around 2.2, as expected from relativistic diffuse shock acceleration, whereas the remaining group requires a harder injected spectrum. The underlying physical mechanisms responsible for different groups may be related to the acceleration process or to the environment properties.

2.4. Search for Gamma Ray Bursts in scaler mode

In scaler mode, the energy threshold for photons is about 1 GeV, lower than the highest energies detected by satellite experiments. Moreover, the modular structure of the ARGO-YBJ detector allowed the collection of data during the different mounting phases. Therefore a search for emission from Gamma Ray Bursts (GRBs) in coincidence with satellite detections started in November 2004, when the Swift satellite was launched [51]. Until February 2013 a sample of 206 GRBs was analysed, 24 of them with known redshift z. This is the largest sample of GRBs investigated with a ground-based detector at high energies. Figure 15 shows the distribution of the resulting significances for all the 206 GRBs (black solid line). No significant excess is found, the largest being 3.52σ for GRB 080727C, with a post-trial chance probability of 4.5×10^{-2}. Since long GRBs (duration T90 > 2 s) typically show a softer spectrum, the same distribution is shown in Figure 15 only for the 27 short GRBs (T90 ≤ 2 s, red dashed area). Also in this case, no significant excess is found, the most significant event being GRB 051114 with 3.37σ and a post-trial chance probability of 1.0×10^{-2}

Besides the coincidence analysis for each GRB, a stacked analysis was carried out in order to search for commom features of all GRBs in time or in phase. In the time analysis the counting rates for all the GRBs were added up in nine windows (Δt = 0.5, 1, 2, 5, 10, 20, 50, 100 and

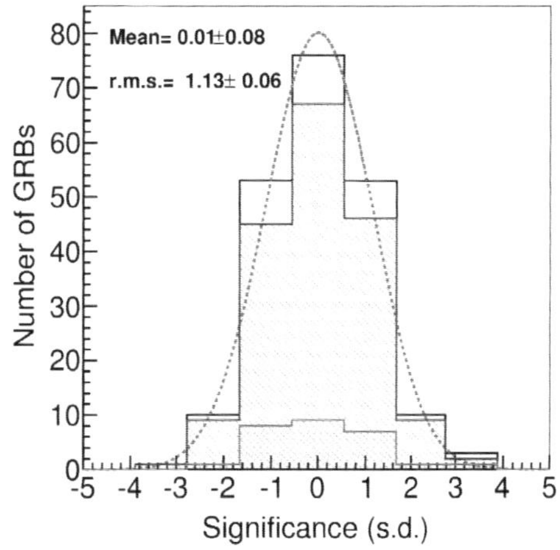

Figure 15. Distribution of the statistical significances of the 206 GRBs with respect to background fluctuations (black solid line) compared with a free Gaussian fit (dotted line, with mean value and r.m.s. given in the plot). The blue and red dashed distributions are for long and short GRBs, respectively.

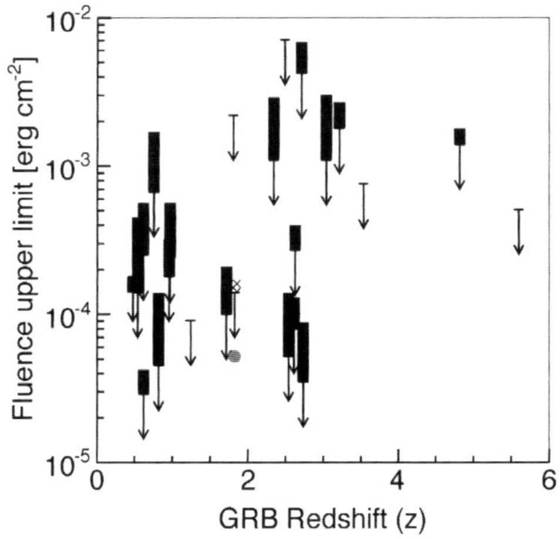

Figure 16. Fluence upper limits in the 1-100 GeV energy range for the 24 GRBs with known redshift, obtained with differential spectral indexes ranging from the value measured by satellites to -2.5 (only this latter case is considered for cutoff power law spectra). The dot shows the fluence extrapolated in the 1-100 GeV interval from the Fermi/LAT observations.

200 s), starting at the trigger time, in order to investigate a possible common duration of the high energy emission. In the phase analysis only the 165 GRBs with duration T90 $\geq$ 5 s were added up, dividing their time length into ten bins each sampling 10% of T90 and scaling their duration to a common phase plot (in this case the lower limit on the GRB duration is due to the minimum interval, 0.5 s, for the scaler mode data acquisition and the number of bins considered). However, there is no evidence of any integral effect in both these analyses.

Since no significant signal was found in the data, for each GRB fluence upper limits in the 1-100 GeV energy range were determined at 99% c.l. assuming two different power law spectra: a) the index measured by satellite detectors in the keV-MeV energy range; b) the conservative differential index -2.5. For case a), when double power law spectral features have been identified, the higher spectral index (i.e., that above the peak in the keV-MeV region of the $E^2 \cdot dN/dE$ spectrum) has been used. For the set of 24 GRBs with known redshift the ranges of upper limits between the values corresponding to the two spectral assumptions are represented by rectangles in Figure 16, while a simple arrow is shown if the low energy spectrum is a cutoff power law, and thus only case b) is considered. These are the only upper limits set at GeV energies for these GRBs. For GRB090902B (which was the GRB in the ARGO-YBJ field of view with the highest photon energy detected) the fluence extrapolated from Fermi/LAT observations in the same energy range is also shown [52]. Only for this GRB the GeV spectral index measured by Fermi/LAT was used and the dashed area in Figure 16 was obtained with a maximum energy of the GRB spectrum ranging from 30 GeV (about the maximum energy measured by Fermi/LAT) to 100 GeV. More results and details about this search can be found in [53].

3. ARGO-YBJ recent highlights in Cosmic Ray Physics

3.1. Measurement of the Cosmic Ray Energy Spectrum

A measurement of the CR primary energy spectrum (all-particle and light nuclei component) is under way with ARGO-YBJ in the wide energy range from few TeV up to about 10 PeV exploiting different approaches:

- *'Digital-Bayes' analysis*, based on the strip multiplicity, i.e. the picture of the EAS provided by the strip/pad system, in the few TeV – 300 TeV energy range. The selection of light elements (i.e. p+He) is based on the particle lateral distribution. The energy is reconstructed, on a statistical basis, by using a bayesian approach [54, 55].
- *'Analog-Bayes' analysis*, based on the RPC charge readout [3], covers the 30 TeV – 10 PeV energy range. The energy is reconstructed (as in the previous analysis), on a statistical basis, by using a bayesian approach [56].
- *'Hybrid measurement'*, carried out by ARGO-YBJ and a wide field of view Cherenkov telescope, in the 100 TeV - 3 PeV region. The selection of (p+He)-originated showers is based on the shape of the Cherenkov image and on the particle density in the core region measured by the ARGO-YBJ central carpet [57, 58].

In the ARGO-YBJ experiment the selection of (p+He)-originated showers is performed not by means of an unfolding procedure after the measurement of electronic and muonic sizes, but on an event-by-event basis exploiting showers topology, i.e. the lateral distribution of charged secondary particles. This approach is made possible by the full coverage of the central carpet, the high segmentation of the read-out and the high altitude location of the experiment that retains the characteristics of showers lateral distribution in the core region.

The all-particle energy spectra measured by ARGO-YBJ by reconstructing showers with three different approaches [56, 59, 60] are shown in the Fig. 17. The statistical uncertainty is shown by the error bars. A systematic uncertainty, due to hadronic interaction models, selection criteria, unfolding algorithms, aperture calculation and energy scale, of $\pm 15\%$ is estimated. The ARGO-YBJ all-particle spectrum clearly shows a knee-like structure at few PeVs in fair agreement with the results obtained by Tibet ASγ [68], IceTop-73 [69], KASCADE [70] and KASCADE-Grande [71] experiments.

3.2. Light component (p+He) energy spectrum of Cosmic Rays

As described in [54, 55], by using the read-out provided by the strip/pad system and applying a selection criterion based on the particle density to quasi-vertical showers ($\theta < 35°$), a sample of events induced by p and He nuclei, with the shower core well inside the ARGO-YBJ central carpet, has been selected. The contamination by heavier nuclei is found negligible. An unfolding technique based on the bayesian approach has been applied to the strip multiplicity distribution in order to obtain the differential energy spectrum of the light component.

The light component (p+He) energy spectrum measured by ARGO-YBJ in the few TeV – 300 TeV energy range is shown in Fig. 18 [54, 55]. Data agree remarkably well with the values obtained by adding up the p and He fluxes measured by CREAM both concerning the total intensities and the spectral index [64]. The value of the spectral index of the power-law fit to the ARGO-YBJ data is -2.64$\pm$0.01 [55]. ARGO-YBJ is the only ground-based experiment that overlaps with the direct measurements for more than two energy decades.

This measurement has been extended to higher energies exploiting an *"hybrid measurement"* with a prototype of the future Wide Field of view Cherenkov Telescope Array (WFCTA) of the LHAASO project [61]. The telescope, located at the south-east corner of the ARGO-YBJ detector, about 78.9 m away from the center of the RPC array, is equipped with 16$\times$16 photomultipliers (PMTs), has a FOV of $14° \times 16°$ with a pixel size of approximately $1°$ [62].

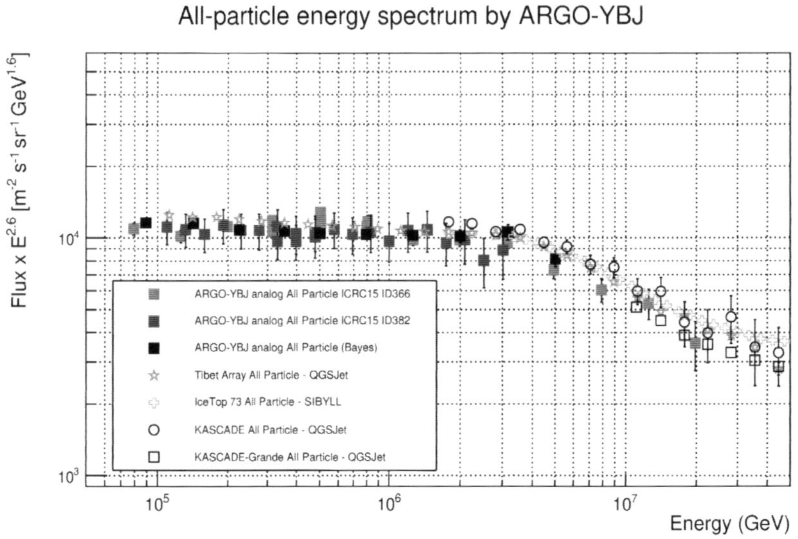

Figure 17. All-particle energy spectrum of primary CRs measured by ARGO-YBJ. Data analyzed with different techniques are compared. The statistical uncertainty is shown by the error bars. For comparison all-particle spectra measured by other experiments (Tibet ASγ [68], IceTop 73 [69], KASCADE [70], KASCADE-Grande [71]) are shown.

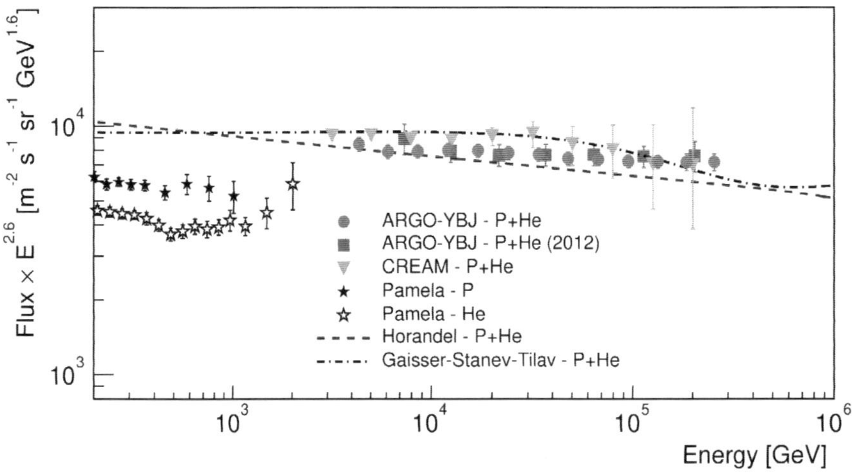

Figure 18. Light (p+He) energy spectrum of primary CRs measured by ARGO-YBJ with the digital readout (strip/pad system) using the full 2008–2012 data sample [55]. The error bars represent the total uncertainty. Previous measurement performed by ARGO–YBJ in a narrower energy range by analyzing a smaller data sample is also reported (blue squares) [54]. The green inverted triangles represent the sum of the proton and helium spectra measured by the CREAM experiment [64]. The proton (stars) and helium (empty stars) spectra measured by the PAMELA experiment [65] are also shown. The light component spectra according to the Gaisser-Stanev-Tilav (dashed–dotted line) [66] and Hörandel (dashed line) [67] models are also shown.

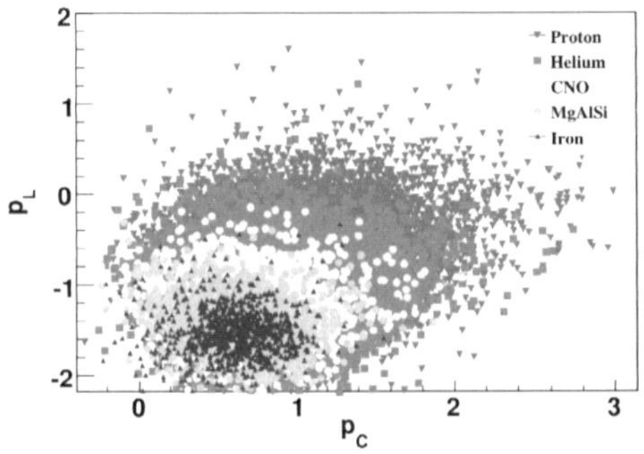

Figure 19. Scatter plot of the parameters p_C and p_L for showers induced by different nuclei. The primary masses have been simulated in the same relative percentage.

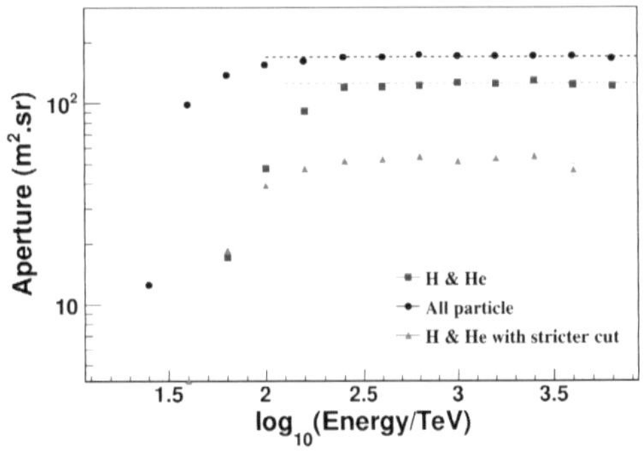

Figure 20. The aperture of the hybrid experiment ARGO-YBJ/WFCTA. Solid circles represent the aperture for all particles, solid squares for the selected p+He events, triangles for the p+He events obtained with stricter cuts for calibration purposes using the low energy part of the spectrum [57].

The idea is to combine in a multiparametric analysis two mass-sensitive parameters: the particle density in the shower core measured by the analog readout of ARGO-YBJ and the shape of the Cherenkov footprint measured by WFCTA [57].

From December 2010 to February 2012, in a total exposure time of 728,000 seconds, the ARGO-YBJ/WFCTA system collected and reconstructed 8218 events above 100 TeV according to the following selection criteria: (1) reconstructed shower core position located well inside ARGO-130, excluding an outer region 2 m large; (2) more than 1000 fired pads on ARGO-130; (3) more than 6 fired pixels in the PMT matrix; (4) a space angle between the incident direction of the shower and the telescope main axis less than 6°. This selection guarantees that the Cherenkov images are fully contained in the FOV, an angular resolution better than 0.3° and a shower core position resolution less than 2 m.

According to the MC simulations, the largest number of particles N_{max} recorded by a RPC in a given shower is a useful parameter to measure the particle density in the shower core region, i.e. within 3 m from the core position. For a given energy, in showers induced by heavy nuclei N_{max} is smaller than in showers induced by light particles. Therefore, N_{max} is a parameter useful to select different primary masses. In addition, N_{max} is proportional to $E_{rec}^{1.44}$, where E_{rec} is the shower primary energy reconstructed using the Cherenkov telescope. We can define a new parameter $p_L = log_{10}(N_{max}) - 1.44 \cdot log_{10}(E_{rec}/TeV)$ by removing the energy dependence [57].

The Cherenkov footprint of a shower can be described by the well-known Hillas parameters [63], i.e. by the width and the length of the image. Older showers which develop higher in the atmosphere, such as iron-induced events, have Cherenkov images more stretched, i.e. narrower and longer, with respect to younger events due to light particles which develop deeper. Therefore, the ratio between the length and the width (L/W) of the Cherenkov image is expected to be another good estimator of the primary elemental composition.

Elongated images can be produced, not only by different nuclei, but also by showers with the core position far away from the telescope, or by energetic showers, due to the elongation of the cascade processes in the atmosphere. Simulations show that the ratio of L/W is nearly proportional to the shower impact parameters R_p, the distance between the telescope and the core position, which must be accurately measured. An accurate determination of the shower

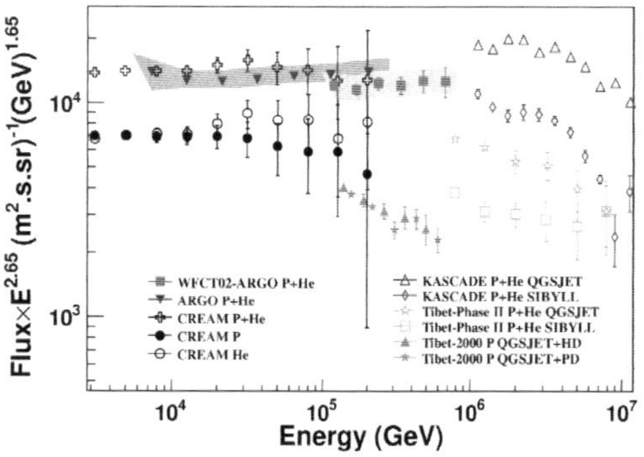

Figure 21. Light component (p+He) energy spectrum of primary CRs measured by ARGO-YBJ/WFCTA hybrid experiment (filled red squares) in the energy range 100 – 700 TeV, compared with other experimental results. The ARGO-YBJ data at lower energies are published in [54].

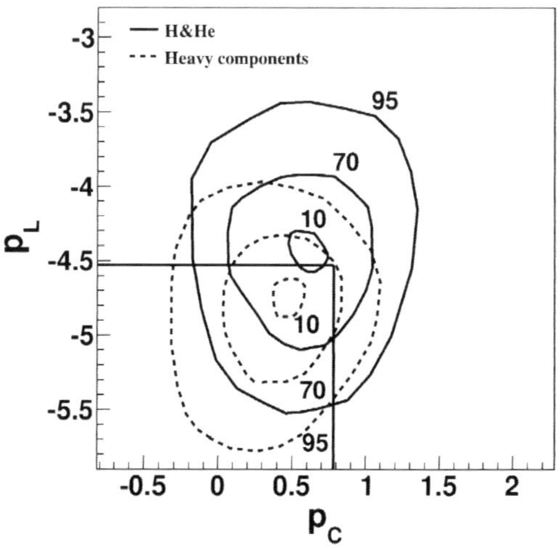

Figure 22. Composition-sensitive parameters p_L and p_C for (p+He) (solid contours) and heavier masses (dashed contours) including 1:1:1 mixing of CNO, MgAlSi, and Iron. The primary energy of the plotted events is in the range 100 TeV – 10 PeV. Numbers on the contours indicate the percentage of contained events.

geometry is crucial for the energy measurement. In fact, the number of photoelectrons collected in the image recorded by the Cherenkov telescope N_{pe} varies dramatically with the impact parameter R_p, because of the rapid falling off of the lateral distribution of the Cherenkov light. Only an accurate measurement of the shower impact parameters R_p, and a good reconstruction of the primary energy allow to disentangle different effects. A shower core position resolution better than 2 m and an angular resolution better than 0.3°, due to the high-granularity of the ARGO-YBJ full coverage carpet, allow to reconstruct the shower primary energy with a resolution of 25%, by using the total number of photoelectrons N_{pe}. The uncertainty in absolute energy scale is estimated about 10%.

Therefore, in order to select the different masses we can define another new parameter $p_C = L/W - 0.0091 \cdot (R_p/1\ m) - 0.14 \cdot log_{10}(E_{rec}/TeV)$ by removing both the effects due to the shower distance and to the energy [57].

The values of these parameters for showers induced by different nuclei are shown in the Fig. 19. The events have been generated assuming a -2.7 spectral index in the energy range 10 TeV – 10 PeV for all the five mass groups (p, He, CNO, MgSi, Fe) investigated. The primary masses have been simulated in the same relative percentage. As can be seen from the figure, a suitable selection in the p_L – p_C space allows to pick out a light composition sample with high purity. In fact, by cutting off the concentrated heavy cluster in the lower-left region in the scatter plot, i.e. $p_L \leq$ -0.91 and $p_C \leq 1.3$, the contamination of nuclei heavier than He is less than 5%. About 30% of H and He survives the selection criteria [57].

The aperture of the ARGO-YBJ/WFCTA system has been estimated using the Horandel model for the primary spectrum [67]. Its value, ~170 m^{2}sr above 100 TeV, shrinks to ~50 m^{2}sr after the selection of the (p+He) component (see Fig. 20). In the sample of 8218 events recorded above 100 TeV by the hybrid system, 1392 showers are selected in the (p+He) sub-sample.

The light component energy spectrum measured by the ARGO-YBJ/WFCTA hybrid system is shown in the Fig. 21 by the filled red squares. A systematic uncertainty in the absolute flux of 15% is shown by the shaded area. The error bars show the statistical errors only. The spectrum can be described by a single power-law with a spectral index of -2.63 ± 0.06 up to about 600 TeV. The absolute flux at 400 TeV is $(1.79 \pm 0.16) \times 10^{-11}$ GeV^{-1} m^{-2} sr^{-1} s^{-1}. This result is consistent for what concern spectral index and absolute flux with the measurements carried out by ARGO-YBJ below 200 TeV and by CREAM. The flux difference is about 10% and can be explained with a difference in the experiments energy scale of ±3.5% [57].

This result is very important to fix the energy scale of the experiment. Below 10 TeV the absolute energy scale of ARGO-YBJ is calibrated at 10% level exploting the westward displacement of the Moon shadow under the effect of the GMF. Above this energy the overposition with CREAM allows to fix the energy scale at few percent level.

3.3. Observation of the knee in the (p+He) energy spectrum

The measurement of the light component energy spectrum has been extended above PeVs exploiting two different approaches.

(1) The ARGO-YBJ/WFCTA hybrid experiment with different selection cuts in the $p_L - p_C$ space [58].

(2) A Bayesian unfolding technique applied to data recorded with the RPC charge readout [56].

(1) ARGO-YBJ/WFCTA hybrid experiment.

In order to extend the measurement of the ARGO-YBJ/WFCTA hybrid experiment to the PeVs, we modified the selection cuts in the p_L-p_C space: events for which $p_L \geq$ -4.53 and $p_C \geq$ 0.78 are rejected (see Fig. 22) [58]. The aperture is a factor 2.4 larger (see Fig. 20). The contamination increase and the purity of the p+He sample below 700 TeV reduces to 93% with respect to 98% estimated with the original cuts. At 1 PeV the contamination is less than 13% increasing to 44% around 6 PeV. About 72% of p+He events survive the selection criteria.

The resulting energy spectrum is shown in Fig. 23 and can be fitted with a broken power-law function

$$J(E) = \begin{cases} J(E_k) \cdot (E/E_k)^{\beta_1} & (E < E_k) \\ J(E_k) \cdot (E/E_k)^{\beta_2} & (E > E_k) \end{cases} \tag{1}$$

with E_k= 700 ± 230 TeV, $J(E_k)=(4.65 \pm 0.27) \times 10^{-12}$ GeV^{-1} m^{-2} s^{-1} sr^{-1}, β_1= -2.56±0.05 and β_2= -3.24±0.36. The relatively large error on the breaking energy E_k is due to the limited statistics. Considering a systematic uncertainty in the absolute energy scale of about 10% (see [58] for a detailed discussion), the systematic uncertainty in E_k is estimated to be ~70 TeV.

The number of expected events in the three energy bins above the knee are 82, 39 and 20, respectively. The difference between the observed number of events and the expectation from a single power law spectrum corresponds to a deficit with a statistical significance of 4.2 standard deviations.

(2) Bayesian unfolding technique. The shower structure in the core region has been deeply investigated thanks to the peculiar characteristics of the ARGO-YBJ detector. According to MC simulations, the truncated shower size (N_8), defined as the number of particles within a radius of 8 m from the shower core, is a good estimator of the primary energy for a given mass and is not affected by bias effects due to the finite detector size [72].

In a shower produced by heavy nuclei a substantial amount of secondary particles is spread further away from the core region. On the contrary, in a shower produced by light elements, the largest amount of particles is concentrated in a small region around the shower core. The ratio between the particle density measured at several distances from the core and the one

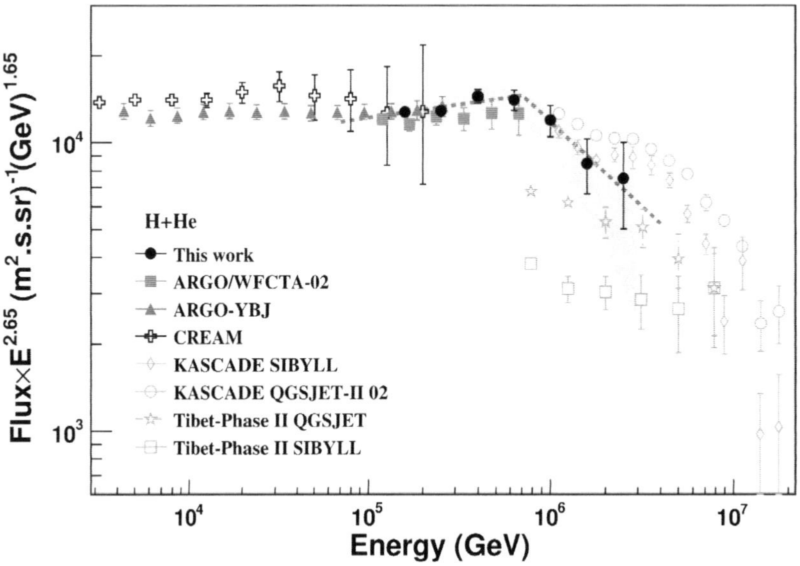

Figure 23. Light (p+He) component energy spectrum measured by the hybrid experiment ARGO-YBJ/WFCTA [58]. The error bar is the statistical error, and the shaded area represents the systematic uncertainty. The (p+He) spectra measured by CREAM [64], ARGO-YBJ [55], Tibet ASγ [68], KASCADE [70] and the hybrid experiment [57] below the knee are shown for comparison.

measured very close to the core can be exploited in order to identify showers produced by light elements. According to Monte Carlo simulations, the parameters $\beta_5 = \rho_5/\rho_0$, where ρ_0 and ρ_5 are respectively the particle density measured in the core region and at 5 m from the core and $\beta_{10} = \rho_{10}/\rho_0$, where ρ_{10} is the particle density measured at 10 m from the core, are sensitive to primary mass. The resolution of the core position reconstruction in ARGO-YBJ determines an uncertainty of about 5% on the measurement of the parameters β_5, β_{10}. As an example in Fig. 24 the distributions of β_5 and β_{10} are reported for data and Monte Carlo events weighted according to the composition model given in [67]. Primaries have been grouped into two mass groups: light (H, He) and heavy (CNO, NeMgSi, Fe). The distributions show a different behaviour depending on the primary mass. The parameters β_5 and β_{10} have been combined in order to estimate the probability that showers have been produced by primaries of different mass.

The determination of the primary energy from the measured quantities is a classical unfolding problem that can be dealt by using an iterative procedure based on the Bayes' theorem [56]. In a probabilistic approach the probability $P(N_8, \beta_5, \beta_{10}|E, A)$ of measuring a shower size N_8 and a certain value of β_5 and β_{10} given a primary energy E and mass A, relates the characteristics of the primary particle to the experimental observables. In a discrete formulation $P(N_8, \beta_5, \beta_{10}|E, A)$ represents the response matrix defined by the following integrals:

$$P(N_8, \beta_5, \beta_{10}|E, A) = \frac{\sum_A \int_{\Delta E_o} dE_o \int_{\Delta\Omega} d\Omega \int dS \; \Phi_M(E_o) \cdot P(N_8, \beta_5, \beta_{10}|E, A)}{\sum_A \int_{\Delta E_o} dE_o \int_{\Delta\Omega} d\Omega \int dS \; \Phi_M(E_o)} \quad (2)$$

where the flux model and the conditional probability are integrated over the primary energy bin E_o^i, the solid angle Ω and the area S projected on a plane perpendicular to the shower direction. The response matrix is normalised to the flux integrated over each bin. The flux model affects

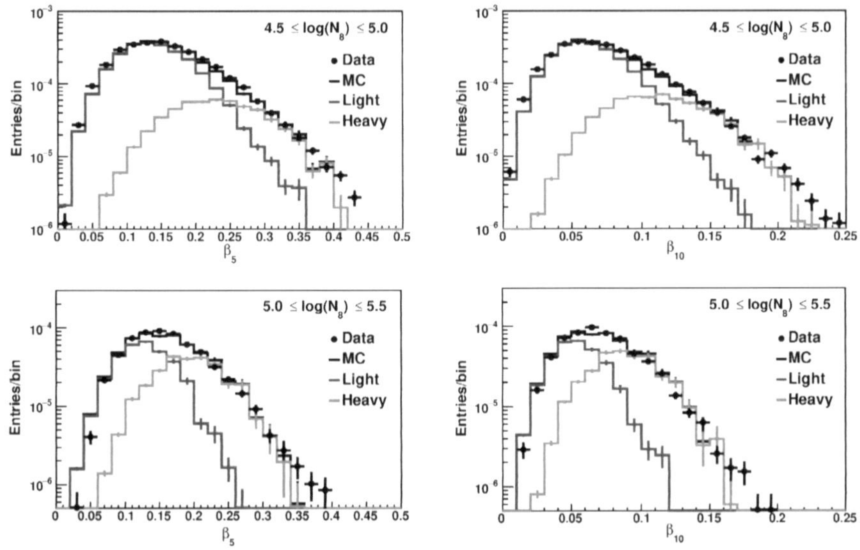

Figure 24. Distribution of β_5 and β_{10} in size bins $4.5 \leqslant \log(N_8) \leqslant 5.0$ and $5.0 \leqslant \log(N_8) \leqslant 5.5$ [56].

weighting events in each bin, but since the energy bins result independent, the flux normalisation cancels out in the response matrix. The quantity $P(N_8, \beta_5, \beta_{10}|E, A)$ can be evaluated by means of a full Monte Carlo simulation of the shower development in the atmosphere and of the detector response giving the primary energy and mass. The number of showers $N(N_{8,i}, \beta_{5,j}, \beta_{10,k})$ in the bin $(i\, j\, k)$ of the measured quantities $(N_8, \beta_5, \beta_{10})$ is therefore related to the number of primaries $N(E_l, A_m)$ by the equation

$$N(E_l, A_m) = \sum_{i,j,k} P(E_l, A_m|N_{8,i}, \beta_{5,j}, \beta_{10,k}) \cdot N(N_{8,i}, \beta_{5,j}, \beta_{10,k}), \tag{3}$$

where $P(E_l, A_m|N_{8,i}, \beta_{5,j}, \beta_{10,k})$ is the probability that a shower of size $N_{8,i}$ and $\beta_{5,j}, \beta_{10,k}$ values has been generated by a nucleus A_m with energy E_l. The probability $P(N_8, \beta_5, \beta_{10}|E, A)$ has to be inverted to obtain the primary energy spectrum.

An iterative procedure based on the Bayes' theorem allows to solve the equation 3 taking into account the bin–to–bin migration due to the fluctuations. Starting from a prior distribution $P_0^{(n)}(E_l, A_m)$ in the nth iteration:

$$P^{(n)}(E_l, A_m|N_{8,i}, \beta_{5,j}, \beta_{10,k}) = \frac{P^{(n)}(N_{8,i}, \beta_{5,j}, \beta_{10,k}|E_l, A_m) P_0^{(n)}(E_l, A_m)}{\sum_{p,q} P^{(n)}(N_{8,i}, \beta_{5,j}, \beta_{10,k}|E_p, A_q) P_0^{(n)}(E_p, A_q)}. \tag{4}$$

An estimate of the energy spectrum for a give mass A_m is therefore obtained from the distribution of the experimental observables

$$N^{(n)}(E_l, A_m) = \sum_{i,j,k} P^{(n)}(E_l, A_m|N_{8,i}, \beta_{5,j}, \beta_{10,k}) \cdot N(N_{8,i}, \beta_{5,j}, \beta_{10,k}), \tag{5}$$

and is used to obtain an updated value of the prior $P_0^{(n)}(E_l, A_m)$. The iterative procedure ends when a variation of the measured flux in two consecutive steps is less than 1%.

In a discrete formulation of the bayesian unfolding approach, the width of the bins has been chosen in order to better evaluate the conditional probabilities namely minimizing the statistical error, reducing bin–to–bin migration effects and stabilising the iterative procedure. Data and Monte Carlo events have been sorted in 20 N_8 logarithmic bins in the range $(3 \div 6)$ and 3 β_5 bins $(0 \div 0.5)$ and 3 β_{10} bins $(0 \div 0.25)$. Monte Carlo events have been sorted in 25 logarithmic energy bins taking into account the energy resolution and 2 mass bins.
The fraction of selected light elements increases with energy and is about 60% above 50 TeV, while contamination is well below 10% over the whole energy range [56].

The energy spectrum of the p+He component measured by the ARGO-YBJ charge read-out is shown in the Fig. 25 where is compared to all measurements carried out by ARGO-YBJ in the energy range TeV - PeV. The p+He measured by CREAM [64] is also shown for comparison. As can be seen, the different analyses show clear evidence of a knee-like structure starting from about 700 TeV. The energy spectrum measured by the analog read-out can be described by the broken power-law formula (1) with these parameters [73]: $E_k = 805 \pm 27$ TeV, $J(E_k)=(2.74\pm0.25)\times10^{-12}$ GeV^{-1} m^{-2} s^{-1} sr^{-1}, $\beta_1 = -2.63\pm0.004$ and $\beta_2 = -3.76\pm0.05$, in good agreement with the results obtained by the ARGO-YBJ/WFCTA measurement.

An overall picture of the energy range TeV - 100 PeV is shown in the Fig. 26 where p+He component and all-particle energy spectra measured by ARGO-YBJ are summarized. Results obtained by other experiments are also reported. For comparison, the parametrization of the light component provided by Horandel [67] is shown by the dashed line.

As can be seen, ARGO-YBJ confirms the observation of the knee in the all-particle energy spectrum at few PeV, but the clear observation of a knee in the p+He energy spectrum below the PeV suggests that the elemental composition in the few PeV range is heavier than Helium, differently from what suggested by the Horandel parametrization [67].

These results demonstrate the possibility of exploring the cosmic ray properties in a wide energy range with a single ground-based detector without exploting the measurement of the muon size, thus reducing the uncertainties due to hadronic interaction models,

4. Conclusions
The ARGO-YBJ detector exploiting the full coverage approach and the high segmentation of the readout imaged the front of atmospheric showers with unprecedented resolution and detail in the wide TeV - PeV energy range.

The physics of Galactic and extragalactic CRs has been studied with a combined measurement of photon- and charged particle-induced showers and a number of important issues in Astroparticle Physics has been investigated (for a summary of the ARGO-YBJ results see [1])

- cosmic ray physics (energy spectrum, elemental composition, anisotropy, p-air and pp cross section measurement) starting from TeVs
- gamma-ray astronomy (galactic and extragalactic) starting from few hundreds GeV
- search from GRBs in the full GeV–TeV energy range
- search for antiproton in the cosmic ray primary flux at TeV energies
- study of the solar and heliosphere physics above GeV.

These achievements represent the accomplishement of the aims wich motivated the proposal and the design of the experiment. Final analysis with the full statistics of the analog data will give new inputs to the hadronic interaction models currently used to describe particle physics and CRs up to the highest energies.

5. Acknowledgements
I am grateful to Conacyt for the support trough the Mesoamerican Center for Theoretically Physics.

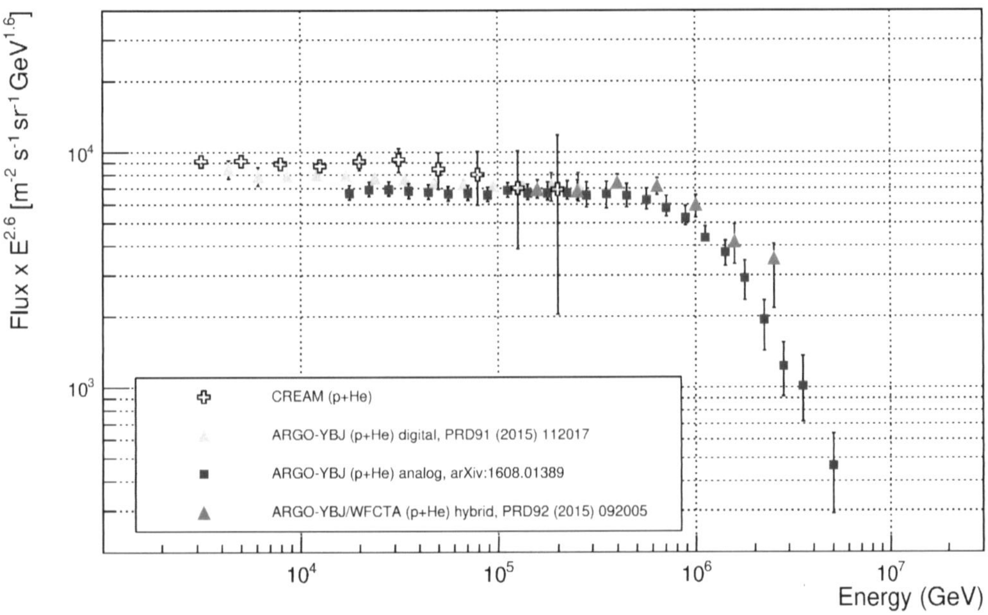

Figure 25. Light (p+He) component energy spectrum of primary CRs measured by ARGO-YBJ. The (p+He) energy spectrum measured by CREAM [64] is shown for comparison. The statistical uncertainty is shown by the error bars.

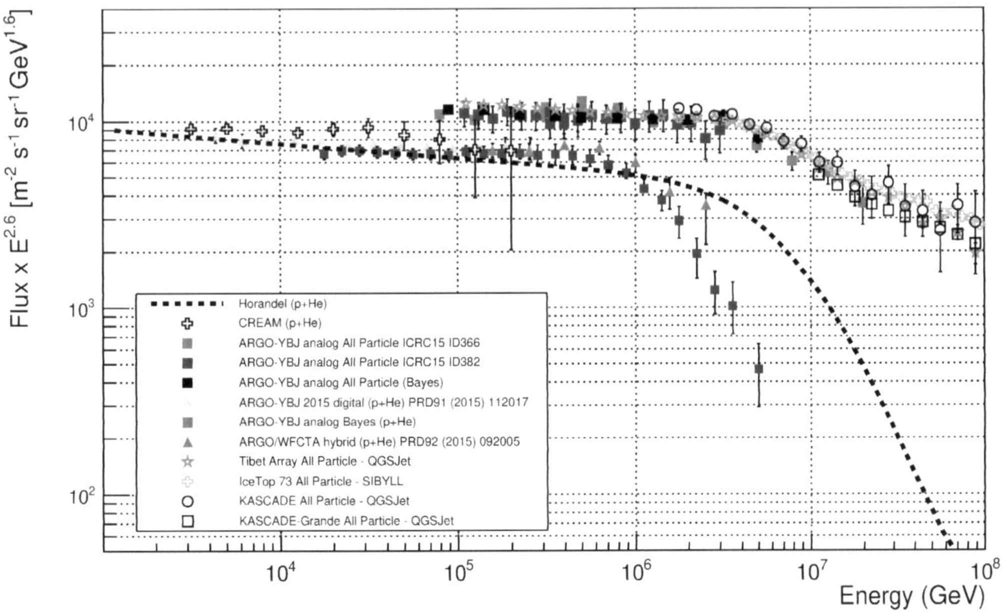

Figure 26. Cosmic ray all–particle and (p+He) energy spectra as measured by ARGO–YBJ compared with other experimental results [64, 68, 69, 70, 71] and with the parametrizations given in [67].

6. References

[1] G. Di Sciascio, *Int. J. Mod. Phys. D* **23** 1430019 (2014).
[2] G. Aielli *et al* (ARGO-YBJ coll.), *Nucl. Instrum. Methods Phys. Res., Sect. A* **562**, 92 (2006).
[3] B. Bartoli *et al* (ARGO-YBJ coll.), *Astropart. Phys.* **67**, 47 (2015).
[4] C. Bacci *et al* (ARGO-YBJ coll.), *Nucl. Instrum. Methods Phys. Res., Sect. A* **443**, 342 (2000).
[5] G. Aielli *et al* (ARGO-YBJ coll.), *Nucl. Instrum. Methods Phys. Res., Sect. A* **608**, 246 (2009).
[6] G. Aielli *et al* (ARGO-YBJ coll.), *Astropart. Phys.* **30**, 287 (2009).
[7] G. Di Sciascio *et al* (ARGO-YBJ coll.), in *Proc. 28th ICRC*, eds. T. Kajita *et al* , (Universal Academy Press, INC, Tokyo, Japan, 2003), **5**, 3015.
[8] G. Di Sciascio *et al* (ARGO-YBJ coll.), in *Proc. 30th ICRC*, eds. R. Caballero et al. (Univ. Nacional Autonoma de Mexico, Merida, 2007), **4**, 123 (arXiv:0710.1945).
[9] G. Aielli *et al* (ARGO-YBJ coll.), *Astropart. Phys.* **30**, 85 (2008).
[10] G. Di Sciascio *et al* (ARGO-YBJ coll.), in *Proc. 29th ICRC*, eds. B. S. Acharya, S. Gupta and S. Tonwar (Tata Institute of Fundamental Research, Mumbai, India, 2005), **6**, 33.
[11] P.J. Huber, *Robust Statistics*, (Wiley, New York, 1981)
[12] S. Vernetto, *Astropart. Phys.* **13**, 75 (2005).
[13] P.K.F. Grieder, *Extensive Air Showers*, (Springer-Verlag, Berlin Heidelberg, 2010) **1**, 277.
[14] B. Bartoli *et al* (ARGO-YBJ coll.), *Phys. Rev. D* **84**, 022003 (2011).
[15] G. Di Sciascio and R. Iuppa, *Nucl. Instrum. Methods Phys. Res., Sect. A* **630**, 301 (2011).
[16] A. A. Abdo *et al* , *Astrophys. J.* **664** L91 (2007).
[17] P. L. Nolan *et al* , *Astrophys. J. Suppl.* **199** 31 (2012).
[18] B. Bartoli *et al* (ARGO-YBJ coll.), *Astrophys. J.* **745** L22 (2012).
[19] F. Aharonian *et al* , å**431** 197 (2005).
[20] J. Albert *et al* , *Astrophys. J.* **675** L25 (2008).
[21] A. A. Abdo *et al* , *Astrophys. J.* **700** L127 (2009).
[22] A. A. Abdo *et al* , *Astrophys. J.* **753** 159 (2012).
[23] M. Ackermann *et al* , *Science* **334** 1103 (2011)
[24] B. Bartoli *et al* (ARGO-YBJ coll.), *Astrophys. J.* **790** 152 (2014)
[25] E. Aliu *et al* , *Astrophys. J.* **770** 93 (2013)
[26] L. O'C. Drury, F. Aharonian & H. J. Völk, å**287** 959 (1994)
[27] A. A. Abdo *et al* , *Astrophys. J.* **658** L33 (2007)
[28] E. Aliu *et al* , *Astrophys. J.* **788** 78 (2014)
[29] B. Bartoli *et al* (ARGO-YBJ coll.), *Astrophys. J.* **760** 110 (2012)
[30] B. Bartoli *et al* (ARGO-YBJ coll.), *Astrophys. J.* **767** 99 (2013)
[31] G. Aielli *et al* (ARGO-YBJ coll.), *Astrophys. J.* **714** L208 (2010)
[32] B. Bartoli *et al* (ARGO-YBJ coll.), *Astrophys. J.* **806** 20 (2015)
[33] S.D. Hunter *et al* , *Astrophys. J.* **481**, 205 (1997).
[34] F.W. Stecker *et al* *Astropart. Phys.* **29**,25 (2008).
[35] A.A. Abdo *et al* , *Phys. Rev. Lett.* **103**, 251101 (2009).
[36] A.A. Abdo *et al* , *Astrophys. J.* **710**, 133 (2010).
[37] R. Atkins *et al* , *Phys. Rev. Lett.* **95**, 251103 (2005).
[38] A. Abdo *et al* , *Astrophys. J.* **658**, L33 (2007).
[39] A.A. Abdo *et al* , *Astrophys. J.* **688**, 1078 (2008)
[40] F. Aharonian *et al* , *Rep. Prog. Phys.* **71**, 096901 (2008).
[41] R. Atkins et al., *Phys. Rev. Lett.* **95** 251103 (2005).
[42] T. Prodanović, B. D. Fields & J. F. Beacom, *Astropart. Phys.* **27** 10 (2007)
[43] M. Ackermann *et al* , *Astrophys. J. Suppl.* **209**, 34 (2013).
[44] D.H. Yan *et al* , *Mont. Not. R. Astron. Soc.* **439**, 2933 (2014).
[45] J. Zhang *et al* , *Astrophys. J.* , 752, 157 (2012).
[46] D.H. Yan *et al* , *Astrophys. J.* , 765,122 (2013).
[47] G.Cao and J.C. Wang, *Publications of the Astronomical Society of Japan* **65**,109 (2013).
[48] Y.P. Peng *et al* , *Mont. Not. R. Astron. Soc.* **442**, 2357 (2014).
[49] Chen, L. 2014, *Astrophys. J.* **788**, 179
[50] B. Bartoli *et al* (ARGO-YBJ coll.), *Astrophys. J. Suppl.* **222** 6 (2016)
[51] G. Aielli *et al* (ARGO-YBJ coll.), *Astrophys. J.* **699** 1281 (2009)

[52] A. A. Abdo *et al* , *Astrophys. J.* **706** L138 (2009)
[53] B. Bartoli *et al* (ARGO-YBJ coll.), *Astrophys. J.* **794** 82 (2014).
[54] B. Bartoli *et al* (ARGO-YBJ coll.), *Phys. Rev. D* **85**, 092005 (2012).
[55] B. Bartoli *et al* (ARGO-YBJ coll.), *Phys. Rev. D* **91**, 112017 (2015).
[56] B. Bartoli *et al* (ARGO-YBJ coll.), preprint arXiv:1608.01389
[57] B. Bartoli *et al* (ARGO-YBJ coll.), *Chinese Phys. C* **38**, 045001 (2014).
[58] B. Bartoli *et al* (ARGO-YBJ coll.), *Phys. Rev. D* **92**, 092005 (2015).
[59] D'Amone A. et al. (ARGO-YBJ coll.), in *Proc. 34th ICRC* (The Hague, 2015) **ID:366**.
[60] M. Iacovacci *et al* (ARGO-YBJ coll.), in *Proc. 34th ICRC* (The Hague, 2015) **ID:382**.
[61] Z. Cao *et al* , *Chinese Phys. C* **34** 249 (2010).
[62] S.S. Zhang *et al* , *Nucl. Instrum. Methods Phys. Res., Sect. A* **A629** 57 (2011).
[63] A.M. Hillas, *Proc. 19th ICRC (La Jolla)* **3** 445 (1985).
[64] Y.S. Yoon *et al* , *Astrophys. J.* **728**, 122 (2011).
[65] O. Adriani *et al* , *Science* 332, 69 (2011).
[66] T. K. Gaisser, T. Stanev, and T. Tilav, *Front. Phys.* **8(6)**, 748 (2013).
[67] J.H. Hörandel, *Astropart. Phys.* **19**, 193 (2003).
[68] M. Amenomori *et al* , *Adv. in Sp. Res.* **47**, 629 (2011).
[69] M.G. Aartsen *et al* , *Phys. Rev. D* **88** 042004 (2013).
[70] T. Antoni *et al* , *Astropart. Phys.* **24**, 1 (2005);
W.D. Apel *et al* , *Astropart. Phys.* **31**, 86 (2009);
W.D. Apel *et al* , *Astropart. Phys.* **47**, 54 (2013).
[71] W.D. Apel *et al* , *Astropart. Phys.* **36**, 183 (2012).
[72] P. Bernardini *et al* , in *Proc. of 34th ICRC* (ICRC 2015), PoS(ICRC2015) *388* (2015).
[73] S. Chen, private communication.

Institute of Physics
Dirac House, Temple Back
Bristol BS1 6BE UK

ISSN: 1742-6588
ISBN 978-1-5108-4421-6

International Workshop on Data Mining in Marketing (DMM 2019)

New York, New York, USA
19 July 2019

ibai - publishing